Study Guide
to accompany

Earth Science
and the Environment
Second Edition

by Thompson & Turk

James R. Albanese
State University of New York
College of Oneonta

SAUNDERS GOLDEN SUNBURST SERIES
Saunders College Publishing
Harcourt Brace College Publishers
Fort Worth Philadelphia San Diego New York Orlando Austin
San Antonio Toronto Montreal London Sydney Tokyo

Portions of this work were published in previous editions.

Printed in the United States of America

ISBN 0-03-021468-8

901 023 7654321

Table of Contents

Chapter 1 Earth Systems

Unit I: Earth Materials and Time
Chapter 2 Minerals
Chapter 3 Rocks
Chapter 4 Geologic Time: A Story in the Rocks

Unit II: Internal Processes
Chapter 5 The Active Earth: Plate Tectonics
Chapter 6 Earthquakes and the Earth's Structure
Chapter 7 Volcanoes and Plutons
Chapter 8 Geologic Structures, Mountain Ranges, and Continents

Unit III: Surface Processes
Chapter 9 Weathering, Soil, and Erosion
Chapter 10 Fresh Water: Streams, Lakes, Ground Water, and Wetlands
Chapter 11 Glaciers and Ice Ages
Chapter 12 Deserts

Unit IV: The Oceans
Chapter 13 Ocean Basins
Chapter 14 Oceans and Coastlines

Unit V: The Atmosphere
Chapter 15 The Earth's Atmosphere
Chapter 16 Weather
Chapter 17 Climate

Unit VI: Human Interactions with Earth Systems
Chapter 18 Climate Change
Chapter 19 Air Pollution
Chapter 20 Water Resources
Chapter 21 Geologic Resources

Unit VII: Astronomy
Chapter 22 Motion in the Heavens
Chapter 23 Planets and their Moons
Chapter 24 Stars, Space, and Galaxies

Chapter 1 Earth Systems

Objectives

After you study this chapter, you should be able to:

- ☐ identify the four spheres of the earth
- ☐ describe the composition of each
- ☐ list forms of interaction between spheres
- ☐ define a system
- ☐ describe types and rates of changes
- ☐ describe the scientific method

Chapter Outline

On a seperate sheet of paper, write a single sentence that summarizes the text for each part of the following outline.

- Earth Science
  - disciplines
    - geology
    - oceanography
    - climatology
    - meteorology
    - physical geography
    - paleontology
    - astronomy
- The Earth's Four Realms
  - Geosphere
    - crust
    - mantle
    - core
  - Hydrosphere
    - ocean water
    - glacial ice
    - ground water
    - surface water
    - atmospheric water
  - Atmosphere
    - gaseous mixture
    - vertical extent
    - functions

- life support
- climate regulation
- Biosphere
  - physical extent
  - human alteration
- Geologic Time
  - origin of concept
  - estimation of magnitude
- The Hot Earth
  - origin of Solar System
  - source of heat
  - formation hypotheses
    - accumulation of rock and metal
    - initial metallic core
  - modern earth
    - magnifications of heat
- Earth Systems
  - definition of system
    - exchange of matter and energy
  - sources of matter
  - sources of energy
  - systems within systems
- Rates of Change in Earth History
  - uniformitarianism (gradualistic)
  - catstrophism
  - threshold and feedback

Checking Vocabulary

Write a definition for each of the Key Terms at the end of this chapter. Check your knowledge of these terms by answering the following questions.

1. The mantle of the earth is part of which realm?
   a. geosphere
   b. hydrosphere
   c. atmosphere
   d. biosphere

2. The innermost layer of the geosphere is called the
   a. crust
   b. mantle
   c. core
   d. lithosphere

3. The ____________ is composed mostly of ocean water.

a. geosphere
b. hydrosphere
c. atmosphere
d. biosphere

4. Sandstone is found in which geospheric layer?

a. crust
b. mantle
c. core
d. hydrosphere

5. Which of the earth's realms is composed of a mixture of gasses?

a. geosphere
b. hydrosphere
c. atmosphere
d. biosphere

6. What part of the hydrosphere penetrates the geosphere?

a. sandstone
b. magma
c. ground water
d. mantle

7. The earth is about how old?

a. 4,600 years old
b. 460,000 years old
c. 4,600,000 years old
d. 4,600,000,000 years old

8. The earth's heat originated from

a. gravitation coalescence
b. impacts of meteorite and comets
c. radioactive decay
d. all of the above

9. Which is an object composed of interacting, related components?

a. a tectonic plate
b. ground water
c. gradualism
d. a system

10. Tectonic plates are found in which part of the geosphere?

a. crust
b. mantle
c. magma
d. core

11. A geologic change which occurs over a long period of time is referred to as
a. gradualism
b. uniformitarianism
c. catastrophism
d. either a. or b.

12. In what order did the realms of the earth form?
a. geosphere, hydrosphere, atmosphere, biosphere
b. hydrosphere, atmosphere, biosphere, geosphere
c. atmosphere, biosphere, geosphere, hydrosphere
d. biosphere, atmosphere, hydrosphere, geosphere

13. Which is the thinnest of the earth's realms?
a. biosphere
b. atmosphere
c. hydrosphere
d. geosphere

14. What characterizes a threshold effect?
a. a small, constant change
b. a large, constant change
c. a small change when a limit is exceeded
d. a large change when a limit is exceeded

15. A small change which causes multiple larger changes is termed a
a. gradualistic change
b. change by uniformitarianism
c. threshold effect
d. feedback mechanism

Reviewing the Facts

Fill in the blank with the word or words which most correctly completes each statement.

1. The __________ is by far the largest of the earth's four realms.

2. The __________ is the outermost layer of the geosphere.

3. The oceans cover ______% of the earth and contain ______% of the hydrosphere.

4. Ninety-nine percent of the __________ is concentrated within 30 kilometers of the earth's crust.

5. The _______ is the zone inhabited by life.

6. The Solar System ________ from a frigid cloud of dust and gas which was rotating in space.

7. The theory explaining that the earth's outer shell is composed of many segments id called the ___________ theory.

8. A system id a combination of _________ and ________ components.

9. ________ energy drives the atmosphere and hydrosphere circulation.

10. The principle that states geologic changes occur over long periods of time is ___________.

11. Catastrophism incorporates that changes in the earth's systems may occur ________.

12. A __________ mechanism occurs when a small initial change in an earth system produces a much large change in a system.

13. The earth is about _________ years old.

14. The biosphere formed about 3.8 billion years age and humanoid life appeared about ______ years ago.

15. The field of ________ studies the composition and relationships between earth's systems.

Developing and Applying Concepts

1. Identify the realm of the earth that each of the following belongs to:

ground water	air	mountains
rocks	glaciers	streams
wind	ocean floor	nitrogen
continents	soil	rain

2. Why is a paleontologist also a geologist?

3. a. How would the earth be different if the processes of precipitation and evaporation din not exist?
   b. Why are precipitation and evaporation important processes for the earth's biosphere?

4. Explain how ocean currents influence where people live.

5. What changes could occur in the hydrosphere if the earth's global temperature increased by 5° C?

6. a. If a significant portion of farmland were to be replaced by shopping malls and parking lots, the ground water supply of the surrounding area could be affected. Explain.
   b. Discuss wether this change would be an example of a threshold effect or a feedback mechanism.

7. Briefly describe the process by which the earth became layered.

8. To develop an understanding of the natural world, scientists use the scientific method. List the steps in this method and describe how they work together.

9. Describe why communication is an important process in understanding the earth.

Thinking Critically about the Environment

1. What observation would you make in order to determine if an industry was polluting the local water supply? What types of information would you be required to obtain from other sources?

2. Describe three ways in which your daily activities impacts each of the earth's four realms. Be specific.

3. State one way in which each of the earth's systems interacts with all the others.

Answers to Questions

Checking Vocabulary

1. a, 2. c, 3. b, 4. a, 5. c, 6. c, 7. d, 8. d, 9. d, 10. a, 11. d, 12. a, 13. a, 14. d, 15. d

Reviewing the Facts

1. geosphere, 2. crust, 3. 71% and 97.5%, 4. atmosphere, 5. biosphere, 6. coalesced, 7. plate tectonic, 8. interrelated, interacting, 9. solar, 10. uniformitarianism, 11. rapidly, 12. feedback, 13. 4.6 billion, 14. 4 million, 15. Earth Science

Developing and Applying Concepts

1.

Geosphere	Hydrosphere	Atmosphere
rocks	ground water	wind
continents	glaciers	air
ocean floor	streams	nitrogen
soil	rain	

mountains

2. A paleontologist studies the remains and traces of ancient life forms found in rocks which are part of the geosphere.

3. a. If the processes of precipitation and evaporation did not exist, there would be no exchange of water between the oceans and the land. All of the hydrosphere would eventually flow into the oceans and stay there. There also would be no clouds increasing the land surface temperatures.
b. These processes are important to the biosphere as without them there would be no life on land. This also would result in the elimination of the organisms which live in freshwater habitats.

4. Ocean currents affect the temperatures of coastal areas, warming some and cooling others. This effect serves to moderate temperatures and make the coastal areas more attractive places in which to live. Ocean currents also provide nutrients for organisms and a means of transportation.

5. The earth's glaciers and ice caps would melt, sea level would rise and the coastal areas would be flooded.

6. a. An increase in the area of solid ground cover would reduce the amount of rain water that may soak into the soil to become ground water. Since ground water is an important source of drinking water, the local populations would suffer.
b. This is a threshold effect as a small change would not impact the population, but over time, when the change was large enough, its effects would become apparent.

7. As the earth formed it heated as the result of gravitation collapse and decay of radioactive elements. Eventually it became hot enough to melt allowing the denser metallic elements to sink toward the center and the less dense elements to rise to the surface. As it cooled, from the outside surface, the layers formed.

8. The scientific method consists of three basic steps:
(1) observation
(2) formation of an hypothesis
(3) testing and development of a theory

These steps work together as a cycle, if the hypothesis fails the testing more observations are taken to reformulate a hypothesis with can withstand the testing and become a theory.

9. communication is an important process in understanding the earth's systems as it increases the opportunities for observations to be made and compared, facilitates the formulation of hypotheses, and allows many individuals to participate in the testing stages of theory development.

Unit I: Earth Materials and Time

Chapter 2 Minerals

Objectives

After you study this chapter, you should be able to:

- ☐ differentiate between a mineral and non-mineral
- ☐ recognize the chemicals represented in a formula
- ☐ describe the chemical properties of a mineral
- ☐ describe the physical properties of a mineral
- ☐ list the groups of rock forming minerals
- ☐ describe silicate and non-silicate minerals

Chapter Outline

On a seperate sheet of paper, write a single sentence that summarizes the text for each part of the following outline.

- Minerals
  - definition
    - naturally occurring
    - solid
    - inorganic
    - chemical composition
    - crystal structure
  - composition
    - elements
      - atoms
        - nucleus
          - protons
          - neutrons
        - electrons
      - ions
        - cation
        - anion
    - compounds
      - chemical bonds
  - crystals
    - crystalline structure

crystal face
physical properties
crystal habit
cleavage
fracture
hardness
Mohs hardness scale
specific gravity
color
streak
luster
rock forming minerals
silicate minerals
silicate tetrahedron
tetrahedra structures
individual
single-chain
double-chain
sheet
framework
carbonate minerals
commercial minerals
ore minerals
industrial minerals
gem minerals

Checking Vocabulary

Write a definition for each of the Key Terms at the end of this chapter. Check your knowledge of these terms by answering the following questions.

1. The manner in which a mineral reflects light is called
    a. streak
    b. color
    c. luster
    d. fluorescence

2. The resistance of a mineral to scratching is called
    a. fracture
    b. hardness
    c. cleavage
    d. streak

3. The basic building blocks of elements are called

a. atoms
b. silica tetrahedra
c. minerals
d. ions

4. A fundamental form of matter that cannot be broken down by ordinary means is a(n)
a. mineral
b. gem
c. element
d. silica tetrahedron

5. Minerals containing silica and oxygen are called
a. elements
b. silicates
c. silica tetrahedra
d. feldspar

6. If a mineral has a specific gravity of four, it weighs
a. 1/4 as much as water
b. the same as water
c. 4 times as much as water
d. 40 times as much as water

7. A silica tetrahedra consists of a silica atom bonded with
a. two oxygen atoms
b. three oxygen atoms
c. four oxygen atoms
d. five oxygen atoms

8. Positively charged ions are called
a. anions
b. cations
c. electrons
d. protons

9. Mohs scale measures
a. luster
b. color
c. specific gravity
d. hardness

10. The nine most common minerals in the earth's crust are called
a. silicate minerals

b. rock-forming minerals
c. accessory minerals
d. ore minerals

11. The type of chemical bond formed when atoms share electrons is
a. ionic
b. metallic
c. covalent
d. van der Waals

12. Negatively charged particles outside the nucleus of an atom are called
a. electrons
b. neutrons
c. protons
d. ions

13. Flat, smooth mineral surfaces which result from breakage are called
a. crystal faces
b. fracture surfaces
c. cleavage planes
d. crystalline structures

14. Minerals are
a. artificial
b. solid
c. organic
d. variable in chemical composition

15. Chemical bonds formed when electrons move freely from one atom to another are
a. ionic
b. metallic
c. covalent
d. van der Waals

Reviewing the Facts

Determine whether the following statements are true or false. If the statement is false change the underlined word so that it is correct.

1. The term conchoidal describes a type of <u>cleavage</u>.

2. <u>All</u> minerals have cleavage.

3. Oxygen and silicon are examples of elements.

4. A mineral, such as mica, which peels off in layers in as example of a sheet silicate.

5. A type of mineral found in great abundance in the earth's crust is garnet.

6. The most abundant group of minerals in the earth's crust are carbonates.

7. Blocky crystals form when silica tetrahedra share oxygen atoms in all directions in a single-chain silicate structure.

8. Physical properties are used to identify minerals.

9. The atomic particle which forms chemical bonds in the proton.

10. The impurities and imperfections of a mineral determine its luster.

11. Common salt is the mineral halite.

12. Metals generally have low specific gravities.

13. Minerals are classified according to the cations they contain.

14. A mineral which measures 10 on the Mohs scale is soft.

15. Color in an unreliable property for mineral identification.

Developing and Applying Concepts

1. Which of the following are not minerals? Explain.

glass	wood	water
steel	halite	ice
emerald	cotton	copper

2. a. Draw a diagram to illustrate the arrangement of protons, neutrons, and electrons in an atom.
   b. What are the similarities between an atom and an ion? What are the differences?

3. The crystal habits of a mineral depend on how small groups of atoms stack together. Make two drawings to show how cubic groups of atoms can form both a long, thin prismatic habit and a cubic one.

4. A mineral's atoms are covalently bonded in one direction and bonded by van der Waals forces in another. Along which direction will cleavage occur? Why?

5. a. What is the hardness of a mineral that will scratch window class but not a steel file?
b. Arrange the following minerals in order of increasing hardness:
Mineral 1 - scratches a penny, does not scratch glass
Mineral 2 - scratches a steel file
Mineral 3 - does not scratch a fingernail

6. a. When ice is dropped into a glass of water it will float. When a piece of gold is dropped into a glass of water, it sinks. Compare the specific gravities of these two substances.
b. Would you expect the most common minerals of the earth's mantle and crust to differ in specific gravity? Explain your answer.

7. Rose quartz (pink), Amethyst (purple), smoky quartz (gray), and citrine (yellow) are all forms of the mineral quartz with the chemical formula SiO_2. Explain how this is possible.

8. Both a shining, silvery mineral with a metallic luster and a dull, red nonmetallic mineral leave reddish-brown marks on a piece of porcelain. Explain.

9. Classify the following minerals into the appropriate groups:

Cu	HgS	$MgAl_2O_4$	Cu_2S
$FeTiO_3Cu_2O$	$AgCl$	TiO_2	
$HgCl$	Ag	Hg	PbS
$PbSO_4$	$AgCuSCuFeS_2$	$MgSO_4.7H_2O$	

10. Of the rock-forming silicate minerals, why are the physical properties of pyroxene and amphibole so similar?

Thinking Critically about the Environment

1. Make a display illustrating the relative amounts of minerals used by people in the United States each year.

2. Make a display of economically important minerals and the products they are used to make.

3. Make a map of your state showing the locations of economically important mineral resources.

Answers to Questions

Checking Vocabulary

1. c, 2. b, 3. a, 4. c, 5. b, 6. c, 7. c, 8. b, 9. d, 10. b, 11. c, 12. a, 13. c, 14. b, 15. b

Reviewing the Facts

1. fracture, 2. some, 3. true, 4. true, 5. feldspar, 6. silicates, 7. framework, 8. true, 9. electron, 10.

color, 11. true, 12. high, 13. anions, 14. hard, 15. true

Developing and Applying Concepts

1. glass - not natural or crystalline
   wood - organic, not crystalline
   water - liquid
   steel - not natural
   cotton - organic

2. a. See figure 2-2
   b. Both atoms and ions have a nucleus containing protons and neutrons. Both also have electrons orbiting the nucleus. Atoms have an equal number of protons and electrons and are neutral in charge. Ions have unequal numbers of protons and electrons and, therefore, have an electrical charge.

3. The cubic crystal will be equidimensional. A prismatic crystal will be longer than wide. See Figures 2- 4 to 2-6.

4. Cleavage will occur along the direction of the van der Waals bonds as this form of bonding is weaker then covalent bonding.

5. a. 6
   b. Mineral 3- hardness less than 2, Mineral 1 - hardness between 3 and 5, Mineral 2 - hardness greater than 6

6. a. The ice has a lower specific gravity than gold.
   b. Yes, you would expect the minerals with the higher specific gravities to be located in the mantle. As the earth solidified the minerals with higher specific gravities sunk and those with lower specific gravities rose toward the crust.

7. Different colors of the same mineral are the results of small impurities, so insignificant in amount, that they do not appear in the chemical formula.

8. The external appearance of a mineral is not always the same as its powder. The outward surface can be effected by the environment (ex. rust).

9.

Cu	$FeTiO_3HgCl$		$PbSO_4$	HgS
Ag	Cu_2O	AgCl	$MgSO_4.7H_2O$	AgCuS
Hg	$MgAl_2O_4$			$CuFeS_2$
	TiO_2			Cu_2S
				PbS

10. Pyroxene is a single-chain silicate and amphibole is a double-chain silicate. The tetrahedra

are arranged in covalently bonded chains in both. The chains are held together with weaker bonds. Therefore, both minerals will have similar cleavage and hardness. Because the elements contained by both are similar, they will have about the same color (black to dark green) and specific gravity.

Chapter 3 Rocks

Objectives

After you study this chapter, you should be able to:

- ☐ name the three groups of rocks
- ☐ discuss the methods of their formation
- ☐ explain the rock cycle
- ☐ list the characteristics of plutonic and volcanic rocks
- ☐ explain how sedimentary rocks form
- ☐ distinguish between mechanisms of metamorphism
- ☐ recognize common rocks of each group

Chapter Outline

On a seperate sheet of paper, write a single sentence that summarizes the text for each part of the following outline.

- Rocks
    - definition
    - formation
        - igneous
        - sedimentary
        - metamorphic
    - decomposition
        - weathering
        - sediments
    - rock cycle
- Igneous rocks
    - magma
        - depth/temperature relationship
    - types
        - extrusive (volcanic)
        - intrusive (plutonic)
    - textures
        - extrusive - small crystals - fine
        - intrusive - large crystals - coarse
    - names
        - mineral composition
            - felsic
            - mafic

            - ultramafic
            - intermediate
    - common forms
        - felsic
            - granite - coarse
            - rhyolite - fine
        - mafic
            - basalt - fine
            - gabbro - fine
        - intermediate
            - andesite - fine
            - diorite - coarse
        - ultramafic
            - peridotite - coarse
- Sedimentary rocks
    - clastic - particle sizes
        - conglomerate - gravel particles
        - sandstone - sand particles
        - siltstone - silt particles
        - shale - clay particles
    - organic - remains of plants & animals
        - chert - silica skeletons
        - coal - plant remains
    - chemical
        - evaporites - ocean water
    - bioclastic
        - carbonate rocks
            - limestone - $CaCO_3$
            - dolostone - $CaMg(CO_3)_2$
        - coquina - larger shells
        - chalk - small shells
- Sedimentary Structures
    - beds / stratification
    - cross-bedding
    - ripple marks
    - mud cracks
    - fossils
- Metamorphic rocks
    - metamorphic grade
        - temperature
        - pressure
        - fluids
    - metamorphic changes

texture
- grains/crystals
- foliations
- slatey cleavages

mineralogy

types of metamorphic processes
- contact metamorphism
- burial metamorphism
- regional metamorphism
  - slate
  - phyllite
  - schist
  - gneiss
  - migmatite
- hydrothermal metamorphism
  - ore deposits

Checking Vocabulary

Write a definition for each of the Key Terms at the end of this chapter. Check your knowledge of these terms by answering the following questions.

1. The abrasion of sediment during transport results in
   a. sorting
   b. rounding
   c. deposition
   d. lithification

2. The metamorphic grade of a rock is most greatly influenced by
   a. geographic location
   b. depth of burial
   c. original mineral composition
   d. intensity of pressure

3. Cementation is an important processes in
   a. deposition
   b. compaction
   c. lithification
   d. foliation

4. Rocks with a high content of magnesium and iron are called
   a. sialic
   b. mafic
   c. porphyritic

d. intermediate

5. Molten rock which is located below the earth's surface is called
   a. magma
   b. sediment
   c. lava
   d. glass

6. Plutonic rocks are
   a. volcanic
   b. finely crystalline
   c. intrusive
   d. formed by quick cooling

7. Heated ground water plays a key role in
   a. contact metamorphism
   b. the development of foliation
   c. lithification
   d. hydrothermal metamorphism

8. Fragments of shells, organic remains, and weathered rock would be found in
   a. chemical sediment
   b. organic sediment
   c. clastic sediment
   d. precipitated sediment

9. The large crystals found in a porphyry are called
   a. phenocrysts
   b. sialic minerals
   c. foliation
   d. sediment

10. The type of rock formed as the result of changes in temperature and pressure is
   a. igneous
   b. sedimentary
   c. metamorphic
   d. chemical

11. The introduction of magma into cold country rock usually results in
   a. regional metamorphism
   b. contact metamorphism
   c. lithification
   d. development of foliation

12. Slaty cleavage is a type of
    a. foliation
    b. rock
    c. sedimentary structure
    d. metamorphism

13. Rock compaction decreases
    a. pore space
    b. grain size
    c. rounding
    d. fissility

14. Extrusive igneous rocks are
    a. plutonic
    b. coarsely crystalline
    c. formed at the earth's surface
    d. formed by cooling slowly

15. The rock cycle theory states that rocks
    a. always stay the way in which they were formed
    b. always change from sedimentary to metamorphic
    c. always change form quickly
    d. may change from any one type to another

Reviewing the Facts

Select one of the two italicized words given in parentheses to correctly complete each of the following statements.

1. As the distance from an intruded magma increases, the grade of metamorphism *(increases, decreases).*

2. According to Bowen's Reaction series, *(sialic, mafic)* rocks form at higher temperatures.

3. Halite is a *(clastic, chemical)* sediment.

4. As distance beneath the earth's's surface increases, temperature *(increased, decreases).*

5. The rock composing most of the ocean floor is *(gabbro, basalt).*

6. Evaporites are a type of *(organic, chemical)* rock.

7. A type of rock that may be classifies as clastic, organic, and chemical is *(limestone, coal).*

8. Changes in texture only occur during the formation of *(marble, schist)*.

9. Obsidian is *(glassy, fine-grained)*.

10. When rocks melt to form magma, it *(expands, contracts)*.

11. Solid-state reactions *(do, do not)* involve melting.

12. Wind can *(round, sort)* clay-sized sediment.

13. The earth's crust was originally *(metamorphic, igneous)*.

14. Clastic rocks are classified according to *(grain size, types of minerals)*.

15. The earth's mantle is probably *(andesite, peridotite)*.

Developing and Applying Concepts

1. Describe the formation of igneous, sedimentary, and metamorphic rocks.

2. Classify each of the following rocks as igneous, sedimentary, or metamorphic:

granite	conglomerate	andesite	hornfels
gneiss	basalt	coquina	obsidian
shale	marble	slate	limestone

3. a. Use the rock cycle to explain each of the following observations:
   Observation 1: fragments of gneiss, sandstone, and granite are found in a conglomerate.
   Observation 2: The distorted shape of an organism is visible in a slate.

   b. Is it possible to tell whether an igneous rock was derived from another major rock type? Explain.

4. a. Use Bowen's Reaction Series to explain why the mineral olivine is not found in granite.

   b. How would you expect the composition of the large crystals of a porphyry to compare with the surrounding matrix? Explain.

5. Compare and contrast the environments of formation of coarsely and finely crystalline igneous rocks.

6. a. Use Figure 3-10 to determine the range of mineral composition of a basalt and a gabbro.

   b. As the amount of sodium-rich plagioclase in a granite or rhyolite increases, what happens to the amounts of potassium feldspar and quartz?

c. Is potassium feldspar ever found in andesite?
d. What three types of minerals can be found in a peridotite?

7. Given that iron and magnesium atoms are heavier than silicon and aluminum atoms, how would you expect the density of mafic rocks to compare with sialic rocks?

8. Imagine you are walking near an old volcano and pick up a rock with relatively large, very dark green crystals in a very fine-grained, medium green matrix. Make a tentative identification of the rock's composition and give the rock a two-part name based on mineral composition and texture.

9. Use Table 3-2 to determine the range of particle sizes found in:
a. claystone
b. siltstone
c. mudstone
d. sandstone
e. conglomerate

10. If a very fine-grained sedimentary rock is fissile, it is called shale (claystone). If it is not, it is called mudstone. Why do you think mudstone is not fissile?

11. Would you expect sand-sized sediment that is derived from a granite to be more well-rounded close to its source rock or farther away? Explain.

12. Classify the following sedimentary rocks as clastic, bioclastic, organic, or chemical:

halite	conglomerate	coquina
coal	gypsum	mudstone
sandstone	chert	chalk

13. The orientation of ripple marks in sedimentary rocks can tell geologists the direction of the water currents which deposited them. What can the orientation of ancient sand dunes indicate?

14. Why is it rare to find fossil remains in most metamorphic rocks?

Thinking Critically about the Environment

1. Visit a local cemetery. Identify the rock type each tombstone is made of, record the date the stone was erected, and evaluate its condition. Which rock types "hold up" better than others? Can you determine why these differences occur?

2. Identify the types of building stones used in your community. Determine the locations where the stone came from. Compare the quantities of each type of building stone and the distances it has traveled.

3. Choose one or more of the following building materials and find out how it was manufactured. In each case, determine where the raw materials come from.

brick
concrete
glass
paint

4. Investigate how salt is mined and determine the effects salt mining and its use has on the environment.

Answers to Questions

Checking Vocabulary

1. b, 2. b, 3.c, 4. b, 5. a, 6. c, 7. d, 8. c, 9. a, 10. c, 11. b, 12. a, 13. a, 14. c, 15. d

Reviewing the Facts

1. decreases, 2. mafic, 3. chemical, 4. increases, 5. basalt, 6. chemical, 7. limestone, 8. marble, 9. glassy, 10. expands, 11. do not, 12. sort, 13. igneous, 14. grain size, 15. peridotite

Developing and Applying Concepts

1. Igneous rocks form when a melt cools and solidifies. Sedimentary rocks from when particles are deposited, compacted, and cemented together; or when minerals are precipitated from water by organic or chemical means. Metamorphic rocks form when a pre-existing rock is changed by temperature and/or pressure.

2.

Igneous	Sedimentary	Metamorphic
granite	shale	gneiss
basalt	conglomerate	marble
andesite	coquina	slate
obsidian	limestone	hornfels

3. a. Observation 1: Sedimentary, metamorphic, and igneous rocks can be weathered and eroded to form sedimentary rocks.
   Observation 2: Sedimentary rocks containing fossils can be changed by temperature and pressure to form metamorphic rocks.
   b. No, because igneous rocks form a melt and all previous combinations of elements into minerals is destroyed.

4. a. Olivine forms at high temperatures, the minerals of a granite are formed at lower

temperatures. If the olivine settles out of the magma chamber, the remaining magma becomes silica-rich and a granite may form, if it does not settle out, it will redissolve as the magma cools further.

b. You would expect the large crystals of the porphyry to be more mafic than the surrounding matrix, as they form first.

5. Coarse grained igneous rocks form beneath the earth's surface where cooling takes place slowly. Finer crystalline igneous rocks form closer to the earth's surface where cooling occurs relatively quickly.

6. a. 5 - 35% olivine
25 - 45% pyroxene
0 - 15% amphibole
20 - 60% calcium plagioclase

b. The amounts of potassium feldspar and quartz decrease

c. no

d. olivine, pyroxene, and calcium-rich plagioclase

7. You would expect the density of mafic rocks to be greater.

8. The rock is probably mafic in composition and could be an andesite porphyry.

9. a. less than 1/256 mm
b. 1/256 - 1/16 mm
c. less than 1/256 - 1/16 mm
d. 1/16 - 2 mm
e. 2 - greater than 256 mm

10. Fissility develops because clay particles are flat. If the rock is composed largely of clay, the flat particles allow planes of weakness to develop along which the rock may split. Silt-sized particles are generally equidimensional and do not lie flat and interfere with clay particle stacking.

11. The farther away a particle is from its source, the more time and opportunities there have been for the particle to have been rounded by abrasion.

12.

Clastic	Bioclastic	Organic	Chemical
sandstone	coquina	coal	halite
conglomerate	chalk	chert	gypsum
mudstone			

13. Ancient wind directions.

14. The process of metamorphism causes elements to recombine to form new minerals, obliterating any fossils.

Chapter 4 Geologic Time

Objectives

After you study this chapter, you should be able to:

•□ distinguish between relative and absolute time

•□ summarize the criteria for relative dating

•□ describe the types of contacts between rocks

•□ discuss the role of fossils in rock correlation

•□ explain the process of radiometric dating

•□ summarize the geological time scale

Chapter Outline

On a seperate sheet of paper, write a single sentence that summarizes the text for each part of the following outline.

- Geologic time measurement
  - relative age
  - absolute age
- Relative geologic time
  - principle of original horizontality
  - principle of superposition
  - principle of cross-cutting relationships
  - gaps in the time record
    - conformable
    - unconformity
    - disconformity
    - angular unconformity
    - non-conformity
  - fossils
    - principle of faunal succession
  - correlation
    - lithologic correlation
    - time correlation
  - index fossil
  - key bed
- Absolute geologic time
  - radioactivity
    - daughter isotope
    - half life

radiometric dating
Geologic column
Geologic time scale
units
eon
era
period
epoch
Hadean Eon
Archean Eon
Proterozoic Eon
Phanerozoic Eon
Paleozoic Era
Mesozoic Era
Cenozoic Era

Checking Vocabulary

Write a definition for each of the Key Terms at the end of this chapter. Check your knowledge of these terms by answering the following questions.

1. To determine that a basalt dike is younger than the horizontally-orientated sedimentary rock it has intruded, one must use
   a. the principle of uniformitarianism
   b. the principle of superposition
   c. the principle of cross-cutting relationships
   d. the principle of original horizontally

2. The longest divisions of the geologic time scale are
   a. eras
   b. periods
   c. epochs
   d. eons

3. A cavity shaped like the outside of an organism is called
   a. internal mold
   b. eternal mold
   c. cast
   d. trace fossil

4. A break in the rock record which represents erosion on non-deposition is called a(n)
   a. nonconformity
   b. unconformity
   c. angular unconformity

d. disconformity

5. The determination of an object's absolute age depends on
   a. the principle of original horizontally
   b. the principle of superposition
   c. the principle of cross-cutting relationships
   d. radiometric age dating

6. The geologic time scale was constructed on the basis of
   a. absolute time
   b. correlation of trace fossils
   c. relative time
   d. radiometric age dating

7. The sudden disappearances of many living species is called
   a. fossilization
   b. carbonization
   c. preservation
   d. extinction

8. The process which preserves organisms by drying is called
   a. carbonization
   b. mineralization
   c. mummification
   d. cast formation

9. The order in which events occurred, without regard to the amount of time separating them, is determined in
   a. absolute time
   b. relative time
   c. geologic time
   d. paleontologic time

10. Precambrian time does not include the
   a. Proterozoic eon
   b. Phanerozoic
   c. Hadean eon
   d. Archean eon

11. The study of ancient life is called
   a. paleontology
   b. correlation
   c. geology

d. earth science

12. The part of the geologic time scale during which organisms first acquired skeletons is the
    a. Mesozoic
    b. Cenozoic
    c. Paleozoic
    d. Precambrian

13. The amount of time it takes for 50% of a radioactive isotope to decay is called
    a. absolute age
    b. relative age
    c. half-life
    d. radioactive decay rate

14. Preserved remains of prehistoric life are called
    a. fossils
    b. isotopes
    c. casts
    d. geologists

15. Atoms which have the same number of protons but different numbers of neutrons are called
    a. isotopes
    b. parents
    c. daughters
    d. different elements

Reviewing the Facts

Fill in the blank with the word or words which most correctly completes each statement.

1. The phrase, "The present is the key to the past," refers to the doctrine of _______________.

2. _______________ dating occurs when events are placed in their proper sequence or order without knowing their absolute ages.

3. The law of superposition is a principle used in ______________dating.

4. The remains or traces of prehistoric life are called _______________.

5. The principle of _______________ states that fossil organisms succeed one another in a definite and determinable order and that any time period can be recognized by its fossil content.

6. Eras are divided into _______________, which can be further divided into _______________.

7. We now live in the _______________ era.

8. The span of time prior to the beginning of the Paleozoic era is termed the _______________.

9. The time required for one-half of the nuclei in a sample of a radio- active isotope to decay is termed _______________.

10. "This dike is 10 million years old." The preceding statement is an example of a(n) _______________ date.

11. Layers of sedimentary rock are said to be _______________ when they are found to have been deposited without interruption.

12. List three different types of unconformities. _______________, _______________, and _______________

13. The process of matching rocks of similar age in different regions is called _______________.

14. The era of "ancient life" is the _______________ era.

15. The era of "middle life" is the _______________ era.

16. The era of "recent life" is the _______________ era.

17. The study of fossils and ancient life is called _______________.

18. Sometimes insects have been preserved in the hardened resin of ancient trees. This material is called _______________.

Developing and Applying Concepts

1. Explain the statement "The present is the key to the past".

2. State the principle on which each of the following inferences is based:
   a. a layer of shale is younger than the layer of sandstone it rests on
   b. a fault is younger than the sedimentary rocks it cuts through
   c. domed sedimentary rocks were once flat lying
   d. horse fossils are found toward the top of the fossil record, so they occurred much later than the fossils in rocks which lie below them
   e. a long period of time is represented by an angular unconformity because the processes which operated when erosion occurred long ago are the same as those which occur today

3. a. What is the half-life of a radioactive isotope if the amount of the parent changes as

given below?

Amount of parent	Amount of time elapsed
400 g	0 hours
300 g	1 hours
200 g	2 hours
150 g	3 hours
100 g	4 hours

b. how much of the isotope will be left after 6 hours?
c. how much time must pass before there are 25 g of the parent left?

4. a. The half-life of a radioactive isotope is 3 million tears, how old is a rock if the parent-daughter ratio is 1:1?
b. How old is the rock if the parent-daughter ratio is 1:7?

5. Why aren't all radioactive age dates accurate?

6. Why might it be useful to use both U - 235 and U - 238 to date a rock?

7. Why are fossils more abundant in the Phanerozoic then in the Precambrian?

8. Identify the process involved in preserving each of the following:
a. thin carbon film of a soft-bodied organism
b. a cavity which preserves the outside form of clam shell
c. the remains of an Egyptian pharaoh
d. the feeding trails of an extinct worm

9. Order the following divisions of geologic time from youngest to oldest: Cambrian, Oligocene, Archean, Triassic, Pennsylvanian, Ordovician, Cretaceous, Jurassic, Miocene, Quaternary

10. Order the following divisions of geologic time from the longest to the shortest: Hadean, Eocene, Triassic, Holocene, Silurian, Precambrian, Cambrian

11. List the following in order of their appearance in the rock record from earliest to latest: flowering plants, brachiopods, horses, bacteria, humans, dinosaurs, land plants, algae

12. a. What theories have been proposed to explain the mass extinctions which have occurred on earth?
b. Why might it seem important to understand the causes of past mass extinctions on earth?

Thinking Critically about the Environment

1. Use the principle of uniformitarianism to explain why it is important to conserve fossil fuels.

2. The ratios of oxygen isotopes in ocean sediments can be used to estimate the extent of glacial buildup in the past. The study of microfossils can indicate whether the ocean surface temperatures were warm or cold and allow an estimate to be made of past global air temperatures. How might information gained from these studies help scientists to evaluate the effects of global warming?

3. It is said that like could not have evolved in the conditions that currently exist on earth. How does the fossil record support that statement?

4. The fossil record contains evidence for catastrophic changes in the earth's environments during its history. How could this evidence be used to anticipate or prevent such changes in the future?

Answers to Questions

Checking Vocabulary

1. c, 2. d, 3. b, 4. b, 5. d, 6. c, 7. d, 8. c, 9. b, 10. b, 11. c, 12. c, 13. c, 14. a, 15. a

Reviewing the Facts

1. uniformitarianism, 2. relative, 3. relative, 4. fossils, 5. superposition, 6. periods, epochs, 7. Cenozoic, 8. Precambrian, 9 . half-life, 10. absolute, 11. conformable, 12. disconformity, nonconformity, angular conformity, 13. correlation, 14. Paleozoic, 15. Mesozoic, 16. Cenozoic, 17. paleontology, 18. amber

Developing and Applying Concepts

1. The precesses which change the earth's surface today are the same processes that occurred in the past. By studying the rates at which these processes currently occur, and their effects, we may gain some insight into the meaning of rocks and the time it took them to form.

2. a. Principle of superposition
b. Principle of cross-cutting relationships
c. Principle of original horizontality
d. Principle of uniformitarianism

3. a. 2 hours
b. 50 grams
c. 8 hours

4. a. 3 million years (one half-life)
b. 9 million years (three half-lives)

5. Parent and/or daughter isotopes may be lost from the rock.

6. The ages obtained from the isotopes of U - 235 and U - 238 could be used to check each other.

7. In the Phanerozoic, organisms evolved shells and skeletons which were more frequently preserved.

8.
   a. carbonization
   b. formation of an external mold
   c. mummification
   d. formation of trace fossils

9. Quaternary, Miocene, Oligocene, Cretaceous, Jurassic, Triassic, Pennsylvanian, Ordovician, Cambrian, Archean

10. Precambrian, Hadean, Cambrian, Triassic, Silurian, Eocene, Holocene

11. Bacteria, algae, brachiopods, land plants, dinosaurs, flowering plants, horses, humans

12.
   a. The meteorite theory, increased volcanic activity, renewed oceanic circulation, and global lowering of sea level, are all theories that have been proposed to explain mass extinctions.
   b. People would like to know what caused mass extinctions so that they can anticipate future events that might drastically change living conditions on Earth. Knowing the causes of past conditions might also help people evaluate the consequences of actions they take to change conditions on earth.

Unit II: Internal Processes

Chapter 5 Plate Tectonics

Objectives

After you study this chapter, you should be able to:

- ☐ describe plate tectonic theory
- ☐ discuss the development of plate tectonic theory
- ☐ list the major types of plate boundaries
- ☐ explain the forces driving the tectonic plates
- ☐ describe the principles of isostatic adjustment

Chapter Outline

On a seperate sheet of paper, write a single sentence that summarizes the text for each part of the following outline.

- Plate Tectonic Theory
- Earth Layers
    - crust
        - continental
        - oceanic
    - mantle
        - lithosphere
        - asthenosphere
    - core
- Plates
    - boundaries
        - divergent
        - convergent
        - transform
    - Divergent plate boundaries
        - spreading center / rift zone
        - mid-oceanic ridge
        - continental rifting
        - rift valley
    - Convergent plate boundaries
        - oceanic / continental crust convergence
        - oceanic / oceanic convergence
        - continental / continental crust convergence

Transform plate boundaries
structure of a tectonic plate
Plate movements
volcanoes
earthquakes
mountain building
migrating contents and oceans
effects on:
hydrosphere
atmosphere
biosphere
movement mechanism
heat engine
mantle plumes
supercontinents
microcontinents
vertical movement
isostacy
isostatic adjustment

Checking Vocabulary

Write a definition for each of the Key Terms at the end of this chapter. Check your knowledge of these terms by answering the following questions.

1. A plate boundary is not a
   a. fault
   b. place where the earth's surface is changed by active forces
   c. site of earthquake origin
   d. place where volcanoes are always found

2. A zone of earthquakes which traces the path of a subducting plate into the mantle is called a(n)
   a. tectonic zone
   b. rift valley
   c. Benioff zone
   d. oceanic trench

3. The idea that one part of the geosphere may "float" on another is called
   a. continental drift
   b. mantle plume theory
   c. continental rifting
   d. isostasy

4. The Mid-Atlantic ridge

a. goes all the way around the earth like a zipper
b. is a spreading canter
c. is a convergent boundary
d. is an island arc

5. The plate tectonics theory states that
a. continental crust "plows through" oceanic crust
b. continental crust "slides over" oceanic crust
c. lithospheric plates move with respect to each other
d. tectonic forces are most active at the center of lithospheric plates

6. Subduction occurs at
a. rift valleys
b. spreading centers
c. transform boundaries
d. oceanic trenches

7. The earth's layers, listed from most to least dense are
a. core, crust, mantle
b. core, mantle, crust
c. mantle core, crust
d. crust, mantle, core

8. The theory of continental drift
a. was based on paleoclimatology
b. suggested that Pangaea existed during the Ordovician period
c. suggested that Laurasia was located south of Pangaea
d. was accepted by the scientific community

9. A lithospheric plate is composed of
a. continental crust only
b. oceanic crust only
c. both continental and oceanic crust
d. either oceanic crust or a combination of oceanic and continental crusts

10. A rift valley does not form at a(n)
a. site of continental rifting
b. divergent plate boundary
c. spreading center
d. oceanic trench

11. A spreading center that winds its way around most of the earth is called
a. Mid-Atlantic ridge

b. Benioff zone
c. mid-ocean ridge
d. San Andreas fault

12. Convection in which hot, less dense material rise in the mantle occurs at
a. convergent and divergent plate boundaries
b. mantle plumes and divergent plate boundaries
c. oceanic trenches and rift valleys
d. subduction zones and transform plate boundaries

13. The part of the earth which flows plastically is the
a. lithosphere
b. asthenosphere
c. continental crust
d. oceanic crust

14. Alternating bands of reversed and normal magnetic polarity are caused by
a. changes in the earth's magnetic field
b. the moon's tidal forces
c. variations in the composition of basaltic magmas
d. the effects of the equipment used to map the ocean floors

15. An island arc is not associated with a(n)
a. subduction zone
b. transform plate boundary
c. oceanic trench
d. Benioff zone

Reviewing the Facts

Determine whether the following statements are true or false. If the statement is false change the underlined word so that it is correct.

1. The mantle is the earth's thinnest and outermost layer.

2. The upper mantle and crust form the lithosphere.

3. The inner core is solid, despite the fact that it is cooler than the outer core.

4. Tectonic plates and lithospheric plates are different materials.

5. A plate boundary is a fracture which separates one plate from the next.

6. A divergent plate boundary may also be referred to as a rift zone.

7. The mid-ocean ridge is a convergent plate boundary.

8. Subduction occurs only at divergent plate boundaries.

9. Divergent plate boundaries occur where two plates slide horizontally past each other.

10. Each tectonic plate is composed of hard, mechanically weak lithospheric rock.

11. All volcanoes are associated with plate boundaries.

12. Many of the earth's largest mountain chains formed at convergent plate boundaries.

13. The location of continents and oceans has no relationship to the movement of lithospheric plates.

14. The earth's internal density is the driving mechanism for plate tectonics.

15. The lithosphere is floating on the asthenosphere in isostatic equilibrium.

Developing and Applying Concepts

1. The earth is composed of three layers of differing rock types and two of differing mechanical properties. List and describe these layers and describe their characteristics.

2. Explain the types of movements that occur at the three different types of plate boundaries.

3. How do spreading centers in continental crust differ from spreading centers in oceanic crust?

4. What determines which plate is subducted when the oceanic crust of one plate converges with the oceanic crust of another?

5. Is it possible that there is convection taking place in the earth's outer core?

6. The convergence of lithospheric plates can produce three different results. List these types of convergence and explain what is produced by each.

7. How would the earth be different if there was no asthenosphere?

8. The distribution of volcanoes and earthquakes is tied to plate tectonic theory. Explain why each of these phenomena is associated with particular types of plate boundaries.

9. How are the ideas of supercontinents and microcontinents related to the fact that the earth's oldest rocks are found on continents?

10. How could you the idea of isostatic adjustment to locate the thinnest area of the earth's crust?

Thinking Critically about the Environment

1. How does the movement of lithospheric plates affect the composition of the gasses in the earth's atmosphere?

2. How is plate tectonic theory related to the cycling of water and carbon between the earth's major realms?

3. How does the changing location of continents and ocean basins affect the earth's climates?

Answers to Questions

Checking Vocabulary

1. d, 2. c, 3. d, 4. b, 5. c, 6. d, 7. b, 8. a, 9. d, 10. d, 11. c, 12. b, 13. b, 14. a, 15. b

Reviewing the Facts

1. crust, 2. true, 3. hotter, 4. the same, 5. true, 6. true, 7. divergent, 8. convergent, 9. transform, 10. strong, 11. earthquakes, 12. true, 13. has a, 14. heat, 15. true

Developing and Applying Concepts

1. The layers of differing rock types are the crust, the thinnest and least dense layer, the mantle, the most voluminous and denser layer, and the core, the layer with the highest density (mostly metallic minerals) and separated into liquid and solid sections. The lithosphere, comprising of the crust and upper mantle is composed of rigid, strong rock, and the asthenosphere, the lower portion of the mantle, is weaker and capable of movement by flowing.

2. The three types of plate boundaries are:
   a. divergent boundaries - two plates move away from each other
   b. convergent boundaries - two plates move toward each other
   c. transform boundaries - two plates slide horizontally past each other.

3. Basaltic ridges form at spreading centers in oceanic crust. Rift valleys form at spreading centers in continental crust.

4. The cooler, older, and therefore denser plate is subducted.

5. Since the outer core is liquid and the inner core is hotter than the mantle, there is the possibility that the heat transfer could drive convection in the outer core.

6. Convergent plate boundaries may be of three forms:

a. Oceanic to continental crust convergence - the denser oceanic crust is subducted under the lower density continental crust,

b. Oceanic to oceanic crust convergence - the older, cooler, denser oceanic crust is subducted under the newer, warmer, lower density oceanic crust,

c. Continental to continental crust convergence - neither low density crust is subducted, producing mountain ranges.

7. If there was no asthenosphere there would be no mountain ranges, less exposed continental crust, much less topographic relief, and no volcanic activity.

8. Volcanoes are associated with divergent and convergent plate boundaries. At divergent boundaries, hot asthenospheric material can escape to the surface, and at convergent boundaries, the subduction of plates causes melting allowing the magma to rise toward the surface. Earthquakes are associated with all three types of plate boundaries.

9. Since continental crust is less dense than oceanic crust it is not subducted. This material floats on the asthenosphere, breaking and joining as the plates move over the earth's surface. Since oceanic crust is subducted and melted, the rocks of the continental crust can last for longer periods of time.

10. Isostatic adjustment allows lithospheric plates to reach an equilibrium between their mass and their elevation. The highest elevation areas have the most mass and sink deeper into the asthenosphere and are therefore the thickest. The lowest elevations have the least mass and are thinner.

Chapter 6 Earthquakes and Structure

Objectives

After you study this chapter, you should be able to:

- ☐ describe the causes of an earthquake
- ☐ list and compare the types of waves produces by an earthquake
- ☐ discuss how earthquakes are measured
- ☐ compare the scales used to describe earthquakes
- ☐ explain the relationship between plate boundary types and earthquakes
- ☐ discuss the prediction of earthquakes

Chapter Outline

On a seperate sheet of paper, write a single sentence that summarizes the text for each part of the following outline.

Earthquake
- components
  - stress
  - elastic deformation
  - plastic deformation
  - brittle fracture
  - fault
- seismic waves
  - focus
  - epicenter
  - body waves
    - primary waves (P)
    - shear, secondary waves (S)
  - surface waves
  - measurement
    - seismograph
    - seismogram
  - measurement scales
    - Mercalli scale
    - Richter scale
    - moment magnitude
  - source location
    - time-travel curve

earthquakes and plate boundaries

- transform boundary
    - strike-slip fault
    - fault creep
    - thrust fault
- convergent boundary - subduction zone
    - Benioff zone
- divergent boundary
    - depths
- plate interior earthquakes

earthquake prediction

- long-term
    - where / when
- short term
    - foreshocks

earthquake damage and mitigation

- rock and soil
    - liquefaction
- construction design
- fire
- landslides
- tsunami
- hazard mitigation

earth's interior

- behavior of seismic waves
- crust-mantle boundary
    - Mohorovicic discontinuity
- mantle structure
    - 600-kilometer discontinuity
- core
    - density
    - magnetic field

Checking Vocabulary

Write a definition for each of the Key Terms at the end of this chapter. Check your knowledge of these terms by answering the following questions.

1. The time-travel curve is not
    a. based in part on the results of atomic testing
    b. used to determine the distance on an epicenter from an observer
    c. used to determine the intensity of an earthquake
    d. based on the differences in arrival time of seismic waves

2. Models of the earth's interior are based on

a. body wave behavior
b. surface wave behavior
c. seismic gap locations
d. earthquake intensity

3. A sudden motion of the earth is called
a. fault creep
b. an earthquake
c. a seismic gap
d. refraction

4. The Richter scale is
a. less reliable than the Mercalli scale
b. used to determine earthquake distances
c. logarithmic
d. based on seismic wavelengths

5. The Mohorovicic discontinuity is the boundary between earth's
a. continental crust and oceanic crust
b. crust and mantle
c. mantle and outer core
d. outer core and inner core

6. Extremely long wavelength, low amplitude waves which result from submarine earthquakes are called
a. tsunamis
b. surface waves
c. L waves
d. seismic waves

7. The type of energy stored when an object is bent under pressure is called
a. friction
b. seismic
c. tectonic
d. elastic

8. The first type of seismic wave to reach an observer is a(n)
a. P wave
b. L wave
c. S wave
d. R Wave

9. The last seismic waves to reach an observer are the

a. secondary waves
b. short waves
c. surface waves
d. slow waves

10. A seismic gap is characterized by
    a. fault creep
    b. a series of small earthquakes
    c. immobility
    d. predictable fore shocks

11. The recording of seismic waves is called a
    a. seismograph
    b. seismologist
    c. seismogram
    d. seismology

12. The bending of seismic waves is called
    a. refraction
    b. reflection
    c. seismology
    d. intensity

13. Earthquake intensity is
    a. based on wave amplitude
    b. the same as earthquake magnitude
    c. based on peoples' experiences
    d. used in the Richter scale

14. Seismic waves originate at the
    a. epicenter
    b. focus
    c. seismograph
    d. seismic gap

15. Small earthquakes which occur before major movement are called
    a. body waves
    b. primary waves
    c. surface waves
    d. foreshocks

Reviewing the Facts

Select one of the two italicized words given in parentheses to correctly complete each of the following statements.

1. Changes in land shape are used for *(long-term, short-term)* earthquake prediction.

2. Force exerted against an object is called *(stress, strain)*.

3. The operation of a seismograph depends on the property of *(gravity, inertia)*.

4. Plate *(boundaries, interiors)* are where earthquakes often occur.

5. Secondary waves *(can, cannot)* travel through liquids.

6. Primary waves travel *(faster, slower)* than secondary waves.

7. Surface waves travel *(faster, slower)* than body waves.

8. To find the location of an earthquake, at least *(two, three)* seismographs must be used.

9. Fault creep involves *(continuous, sudden)* movement along a fault.

10. Many of the earth's strongest earthquakes occur in *(spreading centers, subduction zones)*.

11. The mid-ocean ridge is the site of *(shallow, deep)* focus earthquakes.

12. Long term earthquake prediction can tell *(when, where)* the next earthquake is likely to occur.

13. Construction design plays a *(minor, major)* role in the control of earthquake damage.

14. Sub-sea earthquakes are responsible for the generation of *(volcanic islands, tsunamis)*.

15. The structure of the earth's interior is known *(completely, partially)* from the study of earthquake waves.

Developing and Applying Concepts

1. Explain why a rock may behave elastically.

2. Why do earthquakes generally occur along existing faults?

3. Why do most earthquakes occur at depths within the earth at depths of less than 350 kilometers?

4. Contrast the depths at which earthquakes occur at convergent and divergent plate boundaries.

5. Use Figure 6-11 to answer the following questions:

a. If the difference in the time arrival of P and S waves at a seismic station is 8 minutes, how far is the station from the epicenter of the earthquake?

b. If at 5:30 pm, P waves were recorded at a station located 3,500 km from the epicenter of an earthquake, at what time did the earthquake occur?

6. Why is data needed from three seismic stations to determine the location of an earthquake?

7. a. Why is the Richter scale preferred to the Mercalli scale in describing the strength of an earthquake?

b. Is the strength difference between a magnitude 2 and a magnitude 3 earthquake on the Richter scale the same as the difference between a magnitude 6 and magnitude 7? Explain.

8. Why is a tsunami barely noticeable at sea, but extremely noticeable as it approached a shoreline?

9. Why does construction design play such a large role in the damage produced by earthquakes?

10. List and explain the behaviors of earthquake waves that have allowed the study of the interior of the earth.

Thinking Critically about the Environment

1. If you were going to buy a house in California, what things would you like to know before making the purchase?

2. In earthquake prone areas, there are strict building codes for the construction of new structures. Research these codes and discuss their applicability to older structures.

3. Develop an argument for the non-replacement of buildings destroyed by an earthquake. Expand it to include moving cities to reduce earthquake destruction and deaths.

Answers to Questions

Vocabulary

1. c, 2. a, 3. b, 4. c, 5. b, 6. a, 7. d, 8. a, 9. c, 10. c, 11. c, 12. a, 13. c, 14. b, 15. d

Reviewing the Facts

1. short-term, 2. stress, 3. inertia, 4. boundaries, 5. cannot, 6. faster, 7. slower, 8. three, 9. continuous, 10. subduction zones, 11. shallow, 12. where, 13. major, 14. tsunamis, 15. completely

Developing and Applying Concepts

1. A rock can be called elastic because it can be deformed under pressure and return to its original shape after the pressure is released.

2. It is easier for movement to occur in areas of weakness, as along old faults, than it is for new faults to be created in unbroken rock.

3. Subducted crust warms as it descends into the earth and usually behaves plastically, like the asthenosphere, by the time it reached a depth of 350 kilometers.

4. Earthquakes are shallow at divergent boundaries because the asthenosphere is within 20 to 30 kilometers of the surface. At subduction zones, earthquakes can be much deeper because the subducted plate is cool and dense and encounters friction with the opposing plate as it is forces down.

5. a. 6,500 kilometers
   b. 5:25 pm

6. A seismograph can measure only distance, not direction. The distance measured by a seismograph may be plotted as a circle around it Distance data from at least two other stations, in different locations, may be plotted in the same way. The point at which the three circles intersect indicates the location of the earthquake.

7. a. The Mercalli scale is subjective and based on people's experiences and observations which vary with distance from the earthquake epicenter. The Richter scale is objective and based on data obtained from seismographs.
   b. The differences in magnitudes are not the same because the Richter scale is logarithmic. The difference between a magnitudes 2 and 3 is $10^3 - 10^2 = 1{,}000 - 100 = 900$. The difference between magnitudes 6 and 7 is $10^7 - 10^6 = 10{,}000{,}000 - 1{,}000{,}000 = 9{,}000{,}000$.

8. A tsunami is an ocean wave with an extremely long wavelength and a relatively low wave height. In the open ocean, it travels very fast. As the wave approaches the shoreline, it interacts with the shallow ocean floor and slows, causing the wavelength to decrease and the wave height to increase. A tsunami may increase in height more than 10 times at it approaches shore.

9. Construction design plays a large role in the amount of earthquake damage as most fatalities occur when falling structures crush people. Also as the structures collapse, gas and electrical conduits may be ruptured causing fires. Earthquake resistant designs can prevent these from occurring by allowing the structures to withstand earthquake movements.

10. Seismic waves behave in the following manner:
   a. waves move at a constant velocity in a homogenous medium,

b. wave velocity varies with the density and rigidity of the material it moves through,
c. when a wave encounters a boundary between materials, it may be refracted (bent) and enter the medium, or reflected (bounced back) and return,
d. P waves are compressional allowing them to travel through solids, liquids, and gasses, while S waves are shear waves and can only move through solids.

Chapter 7 Volcanoes and Plutons

Objectives

After you study this chapter, you should be able to:

- ☐ describe the processes by which rocks may melt
- ☐ characterize the two major types of magma
- ☐ list the forms of plutonic rocks
- ☐ list the forms of extrusive rocks
- ☐ discuss the relationship between volcanoes and tectonic plates
- ☐ describe methods of volcanic eruption predictions

Chapter Outline

On a seperate sheet of paper, write a single sentence that summarizes the text for each part of the following outline.

- magma formation
    - temperature
    - pressure
        - pressure-release melting
    - water
    - environments
        - spreading center
        - mantle plume
        - subduction zone
    - partial melting
        - mineral mixture composition
            - granitic
            - andesitic
- magma behavior
    - density changes
    - silica composition
    - water content
- plutons
    - formation and placement
    - batholith
    - stock
    - dike
    - sill
- volcanic rocks

- lava
    - pahoehoe
    - aa
    - columnar joints
- pyroclastic rock
    - cinders
    - volcanic ash
- fissure eruption rock
    - flood basalt
    - lava plateau

volcanoes
- crater
- shield volcano
- cinder cone
- composite cone
- ash flow
    - ash-flow tuff
- caldera

prediction and risk
- regional prediction
- short-term prediction

climate change
- temperatures

Checking Vocabulary

Write a definition for each of the Key Terms at the end of this chapter. Check your knowledge of these terms by answering the following questions.

1. A glowing flow of hot ash and gasses in not a(n)
   a. pyroclastic flow
   b. ash flow
   c. nuée ardente
   d. lava flow

2. Pyroclastic rock which ranges in size from 2 to 64 mm is called
   a. volcanic bombs
   b. tuff
   c. cinders
   d. volcanic ash

3. Pressure-relief melting
   a. raises the melting point of magma
   b. occurs in cooling lavas

c. allows magma to remain liquid at shallow depths
d. results from the contraction of rock

4. The collapsed upper part of a volcanic cone is called a
   a. fissure
   b. caldera
   c. vent
   d. crater

5. A stratovolcano is the same thing as a
   a. cinder cone
   b. shield volcano
   c. composite cone
   d. lava plateau

6. Volcanoes which cover the largest areas are
   a. cinder cones
   b. shield volcanos
   c. composite cones
   d. stratovolcanoes

7. The most violent volcanic activity is associated with
   a. fissure eruptions
   b. cinder cones
   c. shield volcanoes
   d. composite cones

8. Ash which has melted and fused together to form rock is called
   a. pumice
   b. welded tuff
   c. pahoehoe lave
   d. aa lava

9. Lave which cools under water is called
   a. aa lava
   b. pahoehoe lava
   c. pillow basalt
   d. flood basalt

10. A rubbly type of texture is found in
   a. aa lava
   b. pahoehoe lava
   c. pillow basalt

d. flood basalt

11. The largest type of pluton is
    a. dike
    b. stock
    c. sill
    d. batholith

12. Lava cannot flow onto the earth's surface through a
    a. vent
    b. fissure
    c. volcanic neck
    d. crater

13. Flood basalts form
    a. stratovolcanoes
    b. welded tuffs
    c. lava plateaus
    d. shield volcanoes

14. Columnar joints are polygonal columns formed by the cracking of
    a. country rock
    b. shallow basalt sills
    c. batholiths
    d. aa lava

15. A short-lived type of volcano is the
    a. composite cone volcano
    b. stratovolcano
    c. shield volcano
    d. cinder cone volcano

Reviewing the Facts

Determine whether the following statements are true or false. If the statement is false change the underlined word so that it is correct.

1. The smallest pyroclastic particles are termed <u>lapilli</u>.

2. The greater the percentage of silica in magma, the <u>more</u> viscous it is.

3. The process of partial melting produces <u>most</u> magma.

4. <u>All</u> present-day active volcanoes are associated with activity at plate boundaries.

5. Basaltic lava is more viscous than andesitic lava.

6. Partial melting produces a magma having a higher silica content than the parent rock.

7. The largest volcanoes found on earth are of the cinder cone variety.

8. Analysis of samples taken during the eruptions of Hawaiian volcanoes showed that chlorine was the most abundant gas emitted during these events.

9. An active aa lava flow usually advances more rapidly than an active pahoehoe flow.

10. Basaltic magma is less likely to form pyroclasts than andesitic magma.

11. Crater Lake occupies a caldera.

12. A shield volcano is composed largely of pyroclastic fragments.

13. Columnar joints result from the cooling and expansion of a lava flow.

14. Flood basalts are usually intrusive features.

15. Volcanic eruptions have been shown to cause a general cooling of the earth's atmosphere.

Developing and Applying Concepts

1. Contrast the silica and water content of basaltic and granitic magmas.

2. When do columnar joints of a basalt begin to form?

3. List the following in order of increasing size: cinders, volcanic ash, volcanic bombs.

4. How does a lava plateau form?

5. Draw a diagram of a volcano and label the following features: pluton, vent, crater, volcanic cone, country rock.

6. Imagine that you are visiting Crater Lake and notice a relatively small cinder cone surrounded by the water of the caldera's lake. What can you conclude about the age of the cinder cone?

7. Describe the three types of volcanic mountains.

8. Explain the formation and characteristics of a nuée ardente.

9. Describe the characteristics of a volcano which you would look for in order to predict if its next eruption will be quiescent or explosive.

10. What is the mechanism by which volcanic eruptions may cause the cooling of the atmosphere?

Thinking Critically about the Environment

1. If you were thinking of relocation to an area of the country underlain by volcanic rocks, what would you want to know about its history before making your move?

2. Research methods used by scientists to estimate the potential for volcanic eruptions and consequent damages.

3. Investigate where in the Unites States geothermal energy resources might be most easily developed. Make a list of sites already producing geothermal energy and those with some geothermal potential.

4. Discuss some of the ways in which volcanic positively affects the environment.

Answers to Questions

Checking Vocabulary

1. d, 2. c, 3. c, 4. b, 5. c, 6. b, 7. b, 8. b, 9. c, 10. a, 11. d, 12. c, 13. c, 14. b, 15. d

Reviewing the Facts

1. ash, 2. true, 3. true, 4. most, 5. less, 6. true, 7. smallest, 8. water vapor, 9. less, 10. true, 11. true, 12. cinder cone, 13. contraction, 14. extrusive, 15. true

Developing and Applying Concepts

1. Basaltic magma contains about 50% silica, granitic magma contains about 70%. Basaltic magma contains 1 to 2% water, granitic magma contains 10 to 15%.

2. Columnar joints begin to form as a lava cools and contracts. Cracks begin to form when the solid crust is only about ½ centimeter thick.

3. Smallest to largest: volcanic ash, cinders, volcanic bombs.

4. Fluid magma pours from fissures and flows over the surface like flood waters. As successive eruptions occur, layer after layer of rock builds up forming a plateau.

5. Draw a large magma chamber (pluton) that connects to the earth's surface through a narrow vertical shaft (vent). The surrounding rock is the country rock. A conical landform (volcanic cone) is formed in layers around the vent. The crater is the opening for the vent at the top of the volcanic cone.

6. The cinder cone must be younger than the caldera since it is in the lake and not very old in absolute time as cinders erode easily.

7. The three types of volcanic mountains are:

Shield volcanoes - low slopes, formed by fluid magmas,
Cinder cones - small, steep slopes, formed by pyroclastic layers,
Composite cones - formed by alternation layers of pyroclasts and lava flows.

8. After all the magma of an explosive eruption has been ejected, there is no longer any support for the column of ash, rock, and gas above the crater. The column then collapses and behaves as a fluid falling from a great height. The materials then spread rapidly out over the surface, traveling at speeds of up to 200 kilometers per hour. The material is hot and glows brightly as it flows along valleys surrounding the volcano.

9. A volcano composed of basaltic lava flows, such as a shield volcano will probably erupt quiescently as basaltic magmas have lower viscosities and water content. A volcano with a more granitic composition, such as a cinder cone, will probably erupt more violently as granitic magmas have a higher water content and viscosity.

10. Some volcanic eruptions emit large quantities of dust and ash into the atmosphere. These high altitude particles reflect incoming solar radiation and lead to the cooling of the atmosphere.

Chapter 8 Geologic Structures

Objectives

After you study this chapter, you should be able to:

- ☐ list and describe the four types of stress
- ☐ discuss the forms of deformation
- ☐ describe geologic structures
- ☐ explain the relationship between mountain formation and plate boundaries
- ☐ describe the development of the earth's crust

Chapter Outline

On a seperate sheet of paper, write a single sentence that summarizes the text for each part of the following outline.

- Geologic structures
    - stress and deformation
        - confining stress
        - directed stresses
            - compressive stress
            - extensional (tensional) stress
            - shear stress
        - response to stress
            - nature of material
            - temperature
            - pressure
            - time
    - geologic structure
        - fold
            - characteristics
            - types
                - anticline
                - syncline
                - dome
                - basin
        - fault
            - features
                - slip
                - fault zone
                - hanging wall
                - foot wall

        - types
            - normal
                - graben
                - horst
            - reverse
                - thrust
            - strike-slip
    - joints
- geologic structures and plate boundaries
    - divergent boundaries
        - extensional stress
            - normal faults
    - transform boundaries
        - shear stress
            - folds and faults
    - convergent boundaries
        - compressive stress
            - folds
            - reverse and thrust faults
- mountain-building processes
    - orogeny
    - crustal thickening processes
        - subduction zones
        - magmatic activity
        - underthrusting
        - compressive forces
    - mountains at convergent boundaries
        - island arc
            - subduction complex
            - forearc basin
        - continental margin subduction
            - magma generation
            - composition
            - emplacement
        - continental collisions
            - underthrust
- mountain building and earth systems
    - impact of uplift
        - hydrosphere
        - atmosphere
        - biosphere
- origin of continents
    - age of continental rocks

composition
partial melting
horizontal tectonic hypothesis
vertical (plume) tectonic hypothesis

Checking Vocabulary

Write a definition for each of the Key Terms at the end of this chapter. Check your knowledge of these terms by answering the following questions.

1. The backbone of a fold is the
   a. limb
   b. slip
   c. fault plain
   d. axis

2. Any feature produced by the deformation of rock is called a
   a. mountain
   b. graben
   c. fold
   d. geologic structure

3. A highly deformed, sheared, faulted, and metamorphosed mixture of oceanic crust and sea floor sediments is called a(n)
   a. continental suture
   b. island arc
   c. subduction complex
   d. back arc basin

4. The wall which moves down in reverse faults is called the
   a. hanging wall
   b. footwall
   c. limb
   d. slip wall

5. An upward arching fold in which the axis is horizontal is a(n)
   a. syncline
   b. anticline
   c. plunging fold
   d. dome

6. Strike-slip faults
   a. have mainly horizontal movement
   b. have mainly vertical movement

c. are low angle reverse faults
d. are faults where no movement has yet occurred

7. The limb is a feature of a(n)
a. thrust fault
b. anticline
c. dome
d. strike-slip fault

8. If erosion stripped the top off a dome, one would find
a. the oldest rocks were exposed at the center
b. the youngest rocks were exposed at the center
c. a linear pattern of outcropping rock layers
d. both a and c

9. Faults having primarily horizontal movement are called
a. strike-slip faults
b. oblique-slip faults
c. joints
d. normal faults

10. The has been no displacement of rock in a
a. fault
b. fold
c. joint
d. suture

11. An island arc
a. is the final stage of a continental suture
b. consists of the curved, isolated remains of a continent which has been severely eroded
c. consists of isolated mounds of sediment in a subduction complex
d. is a chain of volcanoes next to a trench

12. Which of the following combinations should favor folding rather than faulting?
a. high temperature and low confining pressure
b. low confining pressure and low temperature
c. high confining pressure and low temperature
d. high temperature and high confining pressure

13. Suture zones form at
a. divergent boundaries
b. convergent boundaries
c. transform boundaries

d. Andean margins

14. An elliptical syncline resembling a bowl is called a(n)
    a. basin
    b. anticline
    c. dome
    d. monocline

15. An upwardly displaced block, bounded by normal faults is called a(n)
    a. graben
    b. horst
    c. anticline
    d. mountain

Reviewing the Facts

Select one of the two italicized words given in parentheses to correctly complete each of the following statements.

1. Rock is pulled apart by *(extensional, tensional)* stress.

2. The *(higher, lower)* the temperature, the more likely the rock will behave in a plastic manner.

3. A *(fault, fold)* is a bend in a rock.

4. A fold that arcs *(upward, downward)* is called and anticline.

5. A circular or elliptical syncline is called a *(dome, basin).*

6. Extensional stresses produce *(normal, reverse)* faults.

7. A *(normal, reverse)* fault allows crustal shorting similar to a fold.

8. A thrust fault is a type of *(normal, reverse)* fault that is nearly horizontal.

9. Strike-slip faults are associated with *(divergent, transform)* plate boundaries.

10. The term *(subduction, orogeny)* refers to mountain building processes.

11. Mountain ranges seldom form along *(convergent, transform)* plate boundaries.

12. Along a continent to continent convergence *(subduction, underthrusting)* can lead to the thickening of the earth's crust.

13. Island arcs are more common in the *(Atlantic, Pacific)* ocean, where plate convergence is most common.

14. Mountain ranges *(do, do not)* interact with all of the earth's realms.

15. Continental crust is formed by the *(partial, complete)* melting of rocks as they are subducted.

Developing and Applying Concepts

1. Differentiate between an anticline and a syncline.

2. a. What are the similarities between basins and domes? What are the differences?
   b. Imagine a basin and a dome are eroded flat. Where would the youngest rocks be located in each?

3. The motion along a strike-slip fault may be either to the left or right. Imagine an observer standing on one side of the fault, looking across to the other side. If the opposite side moves to the observer's right, the fault is a right lateral strike-slip or if it moves toward the left, it is a left lateral strike-slip fault. What type of fault is the San Andreas fault?

4. How does the speed at which deformation occurs affect the type of geologic structure produced?

5. Anticlines result from the upward deformation of rock and synclines from downward deformation. Explain why erosion often produces anticlinal valleys and synclinal ridges.

6. Group the following according to whether extensional or compressive tectonic forces were involved:

normal fault	syncline	reverse fault
anticline	graben	horst
dome	thrust fault	basin

7. What do the mid-ocean ridge, the East African rift zone, and the Basin and Range Provence have in common?

8. Describe the formation of a subduction complex and a forearc basin.

9. Explain how continental crust may be formed from mantle rock.

10. Compare and contrast the types of mountains formed by subduction at a continental margin and those formed by a continent to continent collision.

Thinking Critically about the Environment

1. Investigate the interactions between mountain ranges and the hydrosphere, atmosphere, and biosphere.

2. Make a list of ways in which future orogenic activity could affect human populations. What steps could be taken to prepare for these changes?

3. Construct a map showing the current distribution of mountain ranges and populations in the United States. Name the factors which could be the cause of these distributions.

Answers to Questions

Checking Vocabulary

1. d, 2. d, 3. c, 4. b, 5. b, 6. a, 7. b, 8. a, 9. a, 10. c, 11. d, 12. d, 13. b, 14. a, 15. b

Reviewing the Facts

1. extensional, 2. higher, 3. fold, 4.upward, 5. basin, 6. normal, 7. reverse, 8. reverse, 9. transform, 10. orogeny, 11. transform, 12. underthrusting, 13. Pacific, 14. do, 15. partial

Developing and Applying Concepts

1. An anticline is a fold with its limbs extending below its axis, while a syncline has its limbs above the axis of folding.

2. a. Both basins and domes are circular to elliptical in shape. In a basin the rocks in the center have been displaced downward, while in a dome the rocks in the center have moved upward.
   b. The youngest rocks would be found in the center of a basin and around the edge of a dome.

3. The San Andreas is a right lateral strike-slip fault.

4. When deformation occurs quickly, the materials tend to behave brittlely and faulting is the result. Slow deformation allows the rack material to behave more plastically allowing folding to occur.

5. The rock at the top of anticlines is stretched and fractures, causing it to weaken and be easily eroded. The rock at the base of a syncline is compressed and becomes more resistant to erosion.

6. Tectonic forces involved:

extensional	compressive
normal fault	reverse fault
graben	thrust fault

horst	anticline
	syncline
	basin
	dome

7. All are characterized by geologic structures created by stretching forces like those responsible for normal faults and grabens.

8. A subduction complex is formed when sediment from the ocean floor, slices of basalt from the oceanic crust, and pieces of the upper mantle are scraped from a subducted slap and pushed up against an island arc. As more sediment and rock are added, the complex is forces upward and a depression called a forearc basin forms between the complex and the island arc.

9. When the mantle melts, minerals with the lowers melting points become magma first. This partial melting results in the magma having a different mineral composition than the rock from which it formed. The lower density magma could then rise to the surface forming the lower density continental crust.

10. The Andes mountains were formed by subduction at a continental margin. They are characterized by predominantly igneous rocks, with smaller amounts of folded sedimentary rock. The Himalayan mountains were formed by a continent to continent collision and are characterized by a doubled thickness of continental crust sedimentary rock, pushed into large folds and thrust faults.

Unit III: Surface Processes

Chapter 9 Weathering, Soils, and Erosion

Objectives

After you study this chapter, you should be able to:

- ☐ describe the processes of weathering
- ☐ explain the relationship between weathering and erosion
- ☐ describe the development of soil
- ☐ list the types and composition of soils
- ☐ explain the processes of mass wasting
- ☐ list the categories of mass wasting events

Chapter Outline

On a seperate sheet of paper, write a single sentence that summarizes the text for each part of the following outline.

Weathering and erosion
    decomposition of rock
    transport
types of weathering
    mechanical weathering
        processes
            pressure-release fracturing
            frost wedging
            abrasion
            organic activity
            thermal expansion and contraction
    chemical weathering
        processes
            dissolution
            hydrolysis
            oxidation
    combined weathering processes
        salt cracking
        exfoliation
soils
    components

        - regolith
        - organic material - humus
        - water
        - gasses
    - soil profile
        - horizons
            - O horizon
            - A horizon
            - B horizon
            - C horizon
        - leaching
    - formation
        - factors
            - parent rock
            - climate
                - pedocal
                    - caliche
                - pedalfer
                - laterite
                    - bauxite
            - plant growth and decay
            - slope
                - aspect
                - steepness
            - time
        - transport
            - formed elsewhere
- erosion
    - agents
        - streams
        - glaciers
        - wind
    - mass wasting
        - controlling factors
            - slope steepness
            - rock type
            - rock layer orientation
            - unconsolidated material
                - angle of repose
            - water content
            - vegetation
            - earthquakes and volcanoes

- types
  - flow
    - creep
    - mudflow
  - slide
    - slump
    - rockslide
  - fall
- earthquakes and volcanoes
  - triggering mechanisms
- prediction
  - techniques
    - awareness
    - avoidance

Checking Vocabulary

Write a definition for each of the Key Terms at the end of this chapter. Check your knowledge of these terms by answering the following questions.

1. Which one of the following is not true about mechanical weathering?
   a. produces smaller pieces
   b. adds to the effectiveness of chemical weathering
   c. may lead to the formation of talus slopes
   d. changes the rock's mineral composition

2. Chemical weathering would be most effective:
   a. in a warm, dry climate
   b. in a cold, dry climate
   c. in a warm, humid climate
   d. equally in any kind of climate

3. The incorporation and transportation of material by water, wind, or ice describes the process of
   a. weathering
   b. mass wasting
   c. erosion
   d. sheeting

4. The transfer of rock material downslope under the influence of gravity is termed
   a. weathering
   b. mass wasting
   c. hydrolysis
   d. erosion

5. Which of the following is not true of soil
   a. it is the same as regolith
   b. it supports plant growth
   c. it contains organic matter
   d. it contains air and water

6. A soil characteristic of the humid eastern U.S. is
   a. laterite
   b. pedocal
   c. tundra
   d. pedalfer

7. Soil consists of
   a. mineral matter
   b. organic matter
   c. air and water
   d. all of these

8. Which one of the answers below lists soil horizons in the correct order from top to bottom?
   a. A, B, C, O
   b. O, C, B, A
   c. O, A, B, C
   d. C, B, A, O

9. The transfer of rock material downslope under the influence of gravity is termed
   a. weathering
   b. mass wasting
   c. hydrolysis
   d. erosion

10. Which of the following statements best describes slump?
   a. slippage of a mass of material moving as a unit along a curved surface
   b. blocks of bedrock sliding down a slope
   c. a rapid flowage of debris containing a large amount of water
   d. a slow, downhill movement of soil and regolith

11. Which of the following terms is used to describe the way materials move during a mass wasting event?
   a. fall
   b. slide
   c. flow
   d. all of these

12. Alternate wetting and drying or freezing and thawing leads to this
   a. creep
   b. solifluction
   c. mudflow
   d. earthflow

13. An accumulation of angular rock fragments at the base of a steep cliff is termed
   a. caliche
   b. pedocal
   c. talus slope
   d. alluvial fan

14. Bauxite formation is associated with
   a. cool, damp climates
   b. rainy, tropical climates
   c. desert climates
   d. no particular climate

15. A relatively rapid form of flowage that is most common in dry regions
   a. creep
   b. solifluction
   c. mudflow
   d. earthflow

Reviewing the Facts

Fill in the blank with the word or words which most correctly completes each statement.

1. _______________ weathering is accomplished by physical forces that break rock into smaller and smaller pieces without changing the rock's mineral composition.

2. When water in cracks and joints freezes and expands the rock is broken into angular fragments. This process is known as _____________ _______________.

3. _______________ weathering alters the internal structures of minerals by removing and/or adding elements.

4. The general rounding of the corners and edges of angular blocks of rock is termed _______________ _______________.

5. The relative proportions of different particle sizes in a soil is termed _______________.

6. The decayed remains of animal and plant life in soil is termed _______________.

7. Soil consists of mineral matter, ______________, ______________, and _________..

8. A vertical cross-section through a well developed soil reveals layers called ______________.

9. The washing out of fine soil components, usually from the A horizon is known as ______________.

10. The material from which a soil has evolved is called the ___________ material.

11. The downslope movement of rock, regolith, and soil under the direct influence of gravity is called ______________.

12. Loose particles assume a stable slope called the ______________, the steepest angle at which material remains stable.

13. ______________ occurs when material usually saturated with water moves downslope as a viscous fluid.

14. The most rapid form of mass wasting is termed a ______________.

15. The downward slipping of a mass of rock or unconsolidated debris moving as a unit along a curved surface is called ______________.

16. A form of mass wasting that is common to frigid regions underlain by permanently frozen ground is ______________.

Developing and Applying Concepts

1. Identify each of the following as examples of mechanical or chemical weathering:
   - (a) a tree root breaks a rock
   - (b) a crack in a rock widens as water freezes in it
   - (c) feldspar changes to clay
   - (d) quartz grains are loosened from a granite under the influence of gravity
   - (e) the iron in a rock becomes oxidized

2. Explain how chemical weathering processes can increase the rate of mechanical weathering.

3. State the form of mechanical weathering responsible for each of the following rock characteristics and explain how the process occurs:
   - (a) rounded cobbles on a beach
   - (b) huge horizontal cracks in a granite which divide it up in regular "sheets"
   - © pits in rocks by the sea

4. Identify the form of chemical weathering responsible for the following features:
   (a) yellow-brown stains which streak the front of a stone building under the spout of a rain gutter
   (b) the features of a limestone statue in the park which are not as sharp as they used to be

5. Recall that Bowen's Reaction Series states the order in which minerals crystallize as a magma cools. Under the conditions found at the earth's surface, the common rock-forming minerals withstand chemical weathering in the order opposite that in which they crystallize. Knowing this, which of the following would you expect to be more resistant to chemical weathering?
   (a) a basalt dike or granitic country rock
   (b) a quartz sand grain or a hornblende sand grain
   (c) a diorite or a gabbro

6. What could you infer about each of the following if there were abundant grains of pyroxene and hornblende in a slitstone?
   (a) the distance of the source rock
   (b) the rate of tectonic uplift

7. a. What types of weathering breakdown granite? Explain.
   b. Why are most beach sands mainly quartz? Explain.

8. Explain how leaching creates a soil profile.

9. Compare and contrast a pedocal and a pedalfer.

10. List and describe five factors that control mass wasting.

Thinking Critically about the Environment

1. Suppose you are an engineer developing a plan to minimize the likelihood of an earthflow beneath homes built on a steep slope. What measures would you recommend?

2. As a member of the Peace Corps, stationed in the Amazon, it is your job to convince the local populace that cutting down the rainforest to grow crops is not a good plan. How could you do this using the properties of laterite soils and the principles of soil erosion?

3. If you were thinking of buying a house located at the base of a steep, soil covered hill, what geologic characteristics would you want to examine before purchasing?

Answers to Questions

Checking Vocabulary

1. d, 2. c, 3. c, 4. b, 5. a, 6. d, 7. d, 8. c, 9. b, 10. a, 11. d, 12. a, 13. c, 14. b, 15. c

Reviewing the Facts

1. mechanical, 2. frost wedging, 3. chemical, 4. spheroidal weathering, 5. sorting, 6. humus, 7. humus, water, air, 8. horizons, 9. leaching, 10. parent, 11. mass wasting, 12. angle of repose, 13. flow, 14. fall, 15. slump, 16. solifluction

Developing and Applying Concepts

1. (a) mechanical
(b) mechanical
(c) chemical
(d) mechanical
(e) chemical

2. Chemical weathering can speed mechanical weathering by changing a hard mineral to a softer one which is more susceptible to mechanical weathering

3. (a) abrasion - when particles collide, their sharp edges and corners are worn away
(b) pressure release fracturing - when the weight of overlying rock is removed, rocks that were deeply buried expand and develop joints as a result
© salt cracking - salt crystals which form when sea water evaporates exert pressure on the rock and widen spaces between the rock's grains

4. (a) oxidation
(b) dissolution in a naturally formed acid

5. (a) granitic country rock
(b) quartz sand grain
(c) diorite

6. a. The source rock was probably close as pyroxene and hornblende are not very resistant to chemical weathering
b. the rate of uplift was probably very high as these two mafic minerals could only survive in a situation where the rates of erosion and deposition were very high.

7. a. Chemical weathering changes feldspar to clay. The clay will hydrate and expand. Mechanical weathering causes the quartz grains to loosen and fall out of the rock.
b. Quartz in the most resistant to chemical weathering of all the common rock forming minerals.

8. As water moves through any material it may erode and deposit materials. The water picks up nutrients from the decaying organic litter and carries them into the soil. The water dissolves ions from the upper horizons and carries them down into the lower layers, depositing them as the chemical conditions change. As the water continues to flow down, it weathers the bedrock. The

end result is the O horizon, containing abundant humus, the A horizon, the zone of leaching, the B horizon, the zone of accumulation, and a zone of weathering bedrock.

9. A pedocal is formed in arid and semiarid regions by water dissolving ions from the A horizon and evaporating depositing the ions before reaching the bottom of the B horizon. The "cal" in pedocal stands for calcium because heavy calcium deposits, called caliche, form in the B horizon of this type of soil.

A pedalfer is formed in temperate climates. The water which dissolves ions from the A horizon, it carries them into the B horizon, but deposits only iron and aluminum, carrying the calcium ions all the way through the B horizon.

10. (1) steepness of slope - mass wasting occurs more frequently on steeper slopes
(2) rock type and layer orientation - if the rock layers are oriented perpendicular to the slope surface, mass wasting will be less likely than if the layers are parallel to the surface
(3) nature of unconsolidated materials - angular fragments have a higher angle of repose than rounder grains and are less susceptible to mass wasting
(4) water and vegetation - some water will increase the stability of a slope, but too much will add weight to the slope and lubricate the particles. Roots tend to hold unconsolidated materials together
(5) earthquakes and volcanoes - earthquakes provide shocks which may initiate movement on slopes. Volcanic eruptions can melt snow on mountain sides which can cause flows to occur.

Chapter 10 Fresh Water

Objectives

After you study this chapter, you should be able to:

- ☐ describe the hydrologic cycle
- ☐ list the processes by which water cycles
- ☐ explain the factors controlling stream velocity
- ☐ discuss the forms of stream erosion
- ☐ characterize oligotrophic and eutrophic lakes
- ☐ list the factors influencing ground water flow
- ☐ explain ground water erosion
- ☐ describe the impacts of wetlands on the hydrologic cycle

Chapter Outline

On a seperate sheet of paper, write a single sentence that summarizes the text for each part of the following outline.

- Water cycle
    - hydrologic cycle
        - surface waters
            - runoff
            - infiltration
                - ground water
            - evaporation
                - transpiration
- Streams
    - rivers
    - tributaries
    - floods
        - flood plain
    - stream flow
        - velocity
            - gradient
            - discharge
            - channel characteristics
    - stream erosion and sediment transport
        - competence
        - capacity

    - stream load
        - dissolved
        - suspended
        - bed load
            - saltation
    - downcutting
        - base level
            - ultimate base level
    - graded stream
    - sinuosity
        - lateral erosion
        - meanders
        - point bar
        - oxbow lake
        - braided stream
    - alluvial fans
    - delta
        - distributaries
    - drainage basin
- Landscape evolution
    - stream erosion
        - products
    - tectonic activities
        - products
- Floods
    - frequency
    - control
        - artificial levees
        - artificial channels
    - flood plain management
        - techniques

Lakes

- lake cycle
- nutrient balance
    - oligotrophic
    - eutrophic
- salinity
- temperature layers
    - thermocline

Ground water

- characteristics
    - porosity
    - permeability

- water table
  - zone of saturation
  - zone of aeration
  - well
  - recharge
  - aquifer
- movement
  - effluent stream
  - influent stream
- springs
  - perched water table
- artesian aquifer
  - artesian well
- Karst topography
  - cavern
    - stalactite
    - stalagmite
    - column
  - sinkhole
- hot springs
  - heating of ground water
    - geothermal gradient
    - recent volcanism
    - chemical reactions
  - geysers
  - geothermal energy

Wetlands
- characteristics
- biologic impacts

Checking Vocabulary

Write a definition for each of the Key Terms at the end of this chapter. Check your knowledge of these terms by answering the following questions.

1. The release of water vapor to the atmosphere by plants is called
   a. evaporation
   b. transpiration
   c. degassing
   d. infiltration

2. The vertical drop of a stream channel over a certain distance defines
   a. discharge
   b. runoff

c. alluvium
d. gradient

3. Most streams transport the greatest amount of sediment as part of their _______________ load
a. dissolved
b. suspended
c. bed
d. saltation

4. The suspended load of a stream
a. is deposited before the bed load
b. consists primarily of highly soluble substances
c. moves along the bottom of the channel by rolling, sliding, and saltation
d. usually consists of fine sand-, silt-, and clay-sized particles

5. The continuous movement of water from the oceans to the atmosphere, from the atmosphere to the land, and from the land back to the sea is known as
a. evapotranspiration
b. atmospheric cycle
c. lithologic cycle
d. hydrologic cycle

6. Permeable rock strata or sediment that transmit groundwater freely are called
a. perched water tables
b. springs
c. aquicludes
d. aquifers

7. Which one of the following is not true of an artesian well?
a. the water comes from an inclined aquifer
b. the water rises above the level where it was first encountered
c. the water always flows out at the ground surface
d. the aquifer is surrounded by impermeable layers

8. Stalactites
a. hang from the ceiling of a cavern
b. grow up from the floor of a cavern
c. joint floor to ceiling
d. are the same as stalagmites

9. The boundary separating the zone of aeration and the zone of saturation is the
a. aquiclude
b. capillary fringe

c. cone of depression
d. water table

10. Caves most commonly form in
    a. limestone
    b. sandstone
    c. dolomite
    d. granite

11. At a bend in a river, the main erosion is
    a. on the outside of the bend
    b. both outside and inside the bend
    c. on the inside of the bend
    d. at an oxbow lake

12. The boundary between warm and cool temperatures of a lake is called a(n)
    a. inversion
    b. thermocline
    c. water table
    d. meander

13. A lake with a high nutrient supply is termed
    a. thermocline
    b. oligotrophic
    c. eutrophic
    d. karst

14. Geysers generally form from the interactions of
    a. ground water and hot rock
    b. changes in the earth's thermal gradient
    c. chemical reactions
    d. fluctuations in the earth's pressure gradient

15. Wetlands are not characterized by
    a. anaerobic conditions
    b. vegetation adapted to flooding
    c. constant water cover
    d. high biological productivity

Reviewing the Facts

Determine whether the following statements are true or false. If the statement is false, change the underlined word so it is correct.

1. The discharge of a river usually increases downstream.

2. Alluvial fans form when the gradient of a stream is abruptly increased.

3. More water is evaporated from the ocean than is returned to the ocean by precipitation.

4. Ultimate base level is sea level.

5. As its name implies, the water table is always very level (flat).

6. The porosity of a material is an excellent measure of its ability to yield groundwater.

7. For a well to be characterized as being artesian, water must flow freely at the surface.

8. City water systems may be considered artificial artesian systems.

9. The formation of stalactites and stalagmites occurs in the zone of saturation.

10. When a water well is actively pumped, the water table will rise producing a cone-shaped bulge around the well.

11. The source of heat for most hot springs is hot igneous material beneath the surface.

12. Karst topography is most commonly associated with areas underlain by massive sandstone strata.

13. Geothermal power plants have not yet been developed in the United States.

14. A lake with a high nutrient supply is called oligotrophic.

15. Wetland ecosystems may be simply characterized and defined.

Developing and Applying Concepts

1. Describe the hydrologic cycle.

2. What is the difference between runoff and a stream?

3. a. List and explain three factors which control stream velocity.
   b. After an intense spring rain storm, a bridge suddenly collapses. Engineers discover that the base of the bridge supports were undermined. How did this happen?
   c. A stream that flows through a narrow channel bounded on both sides by walls of very hard rock, suddenly opens out onto a broad valley. Describe the changes that will occur in

the stream's shape, velocity, load, and erosive ability.

4. a. why is the discharge of a river greater than the discharge on any one of its tributaries?
b. distinguish between competence and capacity.
c. make an difference that compares the competence and capacity of a river's headwaters in the mountains to its competence and capacity near its mouth.

5. a. how would sediment size along the inner curve of a stream meander compare with that found along the outer side of the same curve?
b. as you walk from the part of the delta nearest the mouth of a river toward the part farthest away, what trend would you notice in sediment size? Explain.

6. If you were to walk from the inside edge of a natural levee out onto the floodplain, what trend would you notice in sediment size?

7. Does a 100 year flood occur once in every 100 years? Explain.

8. What are the advantages and disadvantages of the creation of artificial channels in rivers?

9. Describe the life cycle of a lake.

10. a. why is the withdrawal of water from a well similar to mining a natural resource?
b. why is the prevention of ground water pollution so important?

11. Why is the protection and conservation of wetland areas important to the maintenance of the biosphere?

Thinking Critically about the Environment

1. Investigate the measures which the Army Corps of Engineers has taken to prevent the Mississippi River from shifting into the channel of the Atchafalaya River. Do any of these measures have detrimental side effects on other aspects of the rivers' ecosystems? If so, what are these effects and what do they affect?

2. Living near a river has its benefits and drawbacks. Make a list of the good and bad points of living in this environment.

3. Suppose the creation of a surface water impoundment is being planned in the area in which you live. What aspects of this project would concern you the most and what aspects would be most desirable?

4. Before digging a pond on your property, what parts of the lake life cycle should you consider?

5. Find out where the drinking water in your community comes from. How is the water distributed and what steps are being taken to prevent its pollution?

6. Investigate the applications of geothermal energy. How do you think it would be possible, and cost effective, to use geothermal energy in the area in which you live?

7. How does the extent of wetlands impact on the lives of members of your community? What steps could you take to raise the awareness of their importance?

Answers to Questions

Checking Vocabulary

1. b, 2. d, 3. b, 4. d, 5. d, 6. d, 7. c, 8. a, 9. d, 10. a, 11. a, 12. b, 13. c, 14. a, 15. c

Reviewing the Facts

1. true, 2. decreased, 3. true, 4. true, 5. usually never, 6. permeability, 7. may, 8. true, 9. aeration, 10. fall, 11. true, 12. limestone, 13. have, 14. eutrophic, 15. may not be

Developing and Applying Concepts

1. Water evaporates from the earth's surface and condenses and falls in various forms as precipitation. Some of the precipitation falls on the land and may soak in to become ground water or remain on the surface as lakes, streams, or glaciers. Both ground water and streams flow toward the ocean. Water also can be returned to the atmosphere by plants through the process of transpiration.

2. All streams are runoff because they flow over the land's surface, but all runoff does not flow in channels as streams do.

3. a. the factors, in order of importance, are:
   1) gradient - steeper gradients produce higher velocities
   2) discharge - the greater the volume of water per unit time, the higher the velocity
   3) channel shape - smooth, straight channels have the least friction, producing the greatest velocities

   b. As the discharge of the stream increased, its ability to erode and transport particles increased and it began to cut into its channel, undermining the bridge supports in the process

   c. the water will spread out and the velocity will decrease. The decreased velocity will decrease the stream's ability to erode and transport sediment and deposition will begin.

4. a. a river's volume is the combination of all the tributaries which flow into it. Since each river has many tributaries, its volume is much greater than any single one.

b. stream competence is a measure of the largest particle size it can carry. The capacity of a stream is the total amount of material the stream may carry.
c. At the headwaters of a river it has a high gradient which produces a high velocity giving the water a high competence so large particles may be moved. As the river nears its mouth, the gradient decreases, decreasing competence, but more tributaries enter, increasing discharge and capacity to carry more particles.

5. a. sediment sizes are smaller along the inside of a meander than along the outside because the stream velocity is lower
b. sediment sizes decrease as you move from the upstream side of a delta toward the basin of deposition. As the stream flows toward base level, it loses energy and the largest particles are dropped first.

6. The sediment size would decrease.

7. A 100 year flood is a measure of probability, not a measure of time. The chance of a 100 year flood occurring in any given year is 1 in 100.

8. Artificial channels increase a stream's gradient and velocity, increasing its competence and capacity, resulting in less sediment deposition. The total channel volume, however, is reduced, increasing the possibilities for flooding.

9. A lake is a body of water occupying a depression in the land's surface. As streams carry sediments into the lake its begins to fill, creating, first, a wetland, which supports increased vegetation that can trap more sediment, and finally producing a meadow with a small stream flowing through it.

10. a. Since ground water moves very slowly, it takes a long time for it to be replenished. If the rate of withdrawal exceeds the rate of replenishment, the water will be depleted.
b. ground water is the source of drinking water to over 50% of the United States and the removal of pollutants is much more difficult, if not impossible, than its prevention.

11. Wetlands are one of the most biologically productive ecosystems on earth. Two thirds of the fish and shellfish of the Atlantic consumed by humans use wetlands for a part of their life cycles. one third of the endangered species depend on wetlands for their survival.

Chapter 11 Glaciers and Ice Ages

Objectives

After you study this chapter, you should be able to:

- ☐ explain the formation of a glacier
- ☐ describe the movement of glacial ice
- ☐ list erosional features produced by glaciers
- ☐ describe depositional features produced by glaciers
- ☐ discuss the Pleistocene ice ages

Chapter Outline

On a seperate sheet of paper, write a single sentence that summarizes the text for each part of the following outline.

- Glaciers and Ice Ages
  - Glacial Formation
    - winter snow
    - firn
    - glacial ice
    - types (where formed)
      - alpine glaciers
      - continental ice sheets
- Glacial Movement
  - basal slip
  - plastic flow
  - rates of motion
  - surface feature - crevasses
  - mass balance
    - zone of accumulation
    - zone of ablation
    - snow line
    - terminus
    - icebergs
- Glacial Erosion
  - striations
  - alpine glacier landforms
    - U-shaped valley
    - cirque
    - tarn
    - paternoster lakes

        - horn
        - arête
        - hanging valley
        - fjords
    - continental glacier landforms
- Glacial Deposits
    - drift
        - till
        - stratified drift
    - till landforms
        - erratics
        - moraines
            - end moraine
            - terminal moraine
            - recessional moraine
            - ground moraine
            - lateral moraine
            - medial moraine
        - drumlins
    - stratified drift landforms
        - valley train
        - outwash plain
        - kame
        - esker
        - kettle
- Pleistocene Ice Age
    - timing
    - possible causes
        - volcanic dust
        - variations in earth's orbit
        - feedback mechanisms
    - effects
        - depositional features
            - moraines
            - drumlins
            - soils
        - erosional features
            - lakes
        - sea level changes

Checking Vocabulary

Write a definition for each of the Key Terms at the end of this chapter. Check your knowledge of these terms by answering the following questions.

1. Which one of the following is not true of glaciers?
   a. originate on land
   b. exist only in the Northern Hemisphere
   c. show evidence of past or present flow
   d. form from the recrystallization of snow

2. How does an alpine glacier modify the valley through which it moves?
   a. deepens it
   b. straightens it
   c. widens it
   d. all of these

3. A fiord is
   a. a lake in a cirque
   b. the term used to describe valleys before they were glaciated
   c. a drowned glacial trough
   d. the name given one of the interglacial periods of the Ice Age

4. Drumlins
   a. are composed of outwash
   b. are sinuous ridges formed in tunnels within the glacier
   c. are formed in the zone of fracture
   d. are not found singly but in clusters

5. What feature is created when two lateral moraines join?
   a. esker
   b. medial moraine
   c. arete
   d. col

6. The term "drift"
   a. refers only to deposits made by alpine glaciers
   b. refers only to the sediments that compose moraines
   c. is synonymous with the term "till"
   d. refers to any sediments having a glacial origin

7. Which one of the following is not associated with glacial erosion?
   a. glacial striations
   c. ablation
   b. plucking
   d. abrasion

8. A _________ forms when a block of ice is buried in drift and subsequently melts creating a pit.

a. tarn
b. kame
c. kettle
d. arete

9. Which of the following may have contributed to climatic changes related to the Ice Age?

a. plate movements
b. wobbling of the earth's axis
c. variations in the shape of the earth's orbit
d. all of the above

10. A broad accumulation of stratified drift deposited adjacent to the downstream edge of an end moraine is a

a. kame terrace
b. outwash plain
c. ground moraine
d. esker

11. A long and narrow sinuous ridge composed largely of sand and gravel is a

a. esker
b. arete
c. end moraine
d. kame terrace

12. A streamlined asymmetrical hill composed of till is a

a. arete
b. kame
c. drumlin
d. roche mountonnee

13. Another name for the Ice Age is

a. Pennsylvanian period
b. Pleistocene epoch
c. Miocene epoch
d. Mesozoic era

14. The astronomical theory of Ice Age climatic fluctuations was proposed by

a. Milankovitch
b. de Charpentier
c. Agassiz
d. Venetz

15. A very slow motion of the earth's axis which requires 26,000 years to complete is called
a. precession
b. retrograde motion
c. occultation
d. perturbation

Reviewing the Facts

Select one of the two italicized words given in parentheses to correctly complete each of the following statements.

1. Even when the front of a glacier is retreating, the ice within the glacier is *(advancing, retreating)*.

2. Till is sediment deposited *(directly, marginally)* by a glacier.

3. Except for their positions, terminal and recessional moraines are essentially *(alike, different)*.

4. Glaciers *(can, cannot)* erode below sea level.

5. The Pleistocene epoch is *(one of many, the only)* glacial period for which evidence exists.

6. Glacial valleys are generally *(U-shaped, V-shaped)*.

7. There is evidence for *(five, seven)* glacial periods during earth's history.

8. The orbit of the earth *(cannot, can possibly)* be one of the causes of ice ages.

9. Long Island, N.Y. and Cape Cod, Ma. are composed largely of *(terminal, lateral)* moraines.

10. When continental glaciers melt, the continents *(rise, fall)* isostatically.

11. Glacial striations *(can, cannot)* indicate the direction of movement of a glacier.

12. *(Kettles, Moraines)* are depressions created by the melting of large blocks of ice.

13. Any rock or sediment transported by a glacier is called *(drift, till)*.

14. In the upper zone of glaciers which is too brittle to flow *(drumlins, crevasses)* form.

15. When an ice age occurs, sea level *(falls, rises)*.

Developing and Applying Concepts

1. a. Under what conditions does glacial ice form?
   b. Layering is often visible in glaciers. What factors would make one layer of ice different from the others?
   c. How is glacial ice similar to a metamorphic rock?

2. a. How does glacial movement resemble stream flow?
   b. Occasionally the frozen remains of a person or animal appears at the terminus of a glacier. Make an argument to explain how an organism can die, become frozen in a glacier, remain hidden for thousands of years, and finally appear at the front of a glacier.

3. a. What factors could cause basal slip?
   b. In what types of climates could basal slip be the predominant mechanism of glacial movement?

4. Compare and contrast valley erosion by glaciers and streams.

5. a. Differentiate between till and stratified drift.
   b. group the following as till or stratified drift:
      esker, kame, drumlin, moraine, outwash plain

6. a. When in earth's history have ice ages occurred?
   b. The amount of time between two ice ages is 1.7 billion years. There were from 100 million to 250 million years between the later ice ages. Do you think that ice ages were just less frequent in early earth history?

7. a. How did the last North American continental glacier affect each of the following?
      1) the Great Lakes
      2) Long Island
      3) Northern Great Plains
   b. Can a glacier erode while it is retreating? Explain.

8. a. What would happen to the surfaces of Greenland and Antarctica if their ice sheets melted?
   b. Draw a profile of the land's surface beneath the Greenland ice sheet. Assume the glacier is thickest in the middle and the land was level before glaciation.
   c. How would you respond to the statement "glaciers always move down hill"?

9. What evidence is there that one body of land can be affected by the isostatic rebound on another area?

10. What is a fjord and how are they formed?

Thinking Critically about the Environment

1. Do you think there is a relationship between the rate at which a glacier retreats and the quality of the aquifer it leaves behind?

2. In planning for future development, how wound you use land which was deposited as an outwash plain versus land which is composed of till? What are the possibilities for the development of ground water resources of both?

3. How could you use your knowledge of glacial erosion and deposition in the search for geologic resources? What types of resources could glaciation concentrate?

Answers to Questions

Checking Vocabulary

1. b, 2. d, 3. c, 4. d, 5. b, 6. d, 7. c, 8. c, 9. d, 10. b, 11. a, 12. c, 13. d, 14. a, 15. a

Reviewing the Facts

1. advancing, 2. directly, 3. alike, 4. can, 5. one of many, 6. U-shaped, 7. seven, 8. can possibly, 9. terminal, 10. rise, 11. can, 12. kettles, 13. drift, 14. crevasses, 15. falls

Developing and Applying Concepts

1. a. Glacial ice begins when snow deposited during one winter does not melt during the summer. As the snow accumulates it becomes compacted and glacial ice forms.
b. Variations in the concentrations of air bubbles, water bubbles, dust, and rock fragments causes the ice layers to stand out from one another.
c. Glacial ice is like a metamorphic rock because the ice crystals form as the result of snow being buried and subjected to high pressures. This results in recrystallization without melting.

2. a. Both glacial ice and stream waters move faster in the center of their channels, or valleys than at the edges because of friction.
b. An organism can fall into a crevasse. It would then become incorporated into the ice and flow toward the terminus. At the terminus the glacial ice would melt, revealing the preserved organism.

3. a. Basal slip occurs when meltwater accumulates at the bottom of a glacier and reduces friction there. This can happen when:
    1) the glacier moves over a warm section of the earth's crust
    2) friction produces heat
    3) melting occurs as the result of the pressure of overlying ice
    4) surface meltwater seeps down from the surface.

4. Both streams and glaciers change the shapes of their valleys. Streams produce V-shaped valleys as they erode and flood plains as they deposit. Since glaciers exert pressure in both lateral and downward directions at once, they produce U-shaped valleys.

5. a. Till is unsorted and stratified drift is sorted.
   b.

Till	Stratified drift
drumlin	esker
moraine	kame
	outwash plain

6. a. Earth experienced ice ages approximately 2.3 billion years age, 600 million years ago, 500 million years ago, 250 million years ago, and 2 million years ago.
   b. It is more difficult to find evidence of glaciations in older rocks because they have been metamorphosed or eroded.

7. a. 1) the Great Lakes were scoured and deepened by the continental ice sheets.
      2) Parts of Long Island were deposited as terminal moraines.
      3) Outwash and ground moraines form the fertile soils of the northern Great Plains.
   b. Glacial ice is always moving forward, even if the end of the glacier is retreating.

8. a. The surface would undergo isostatic rebound and rise.
   b. The land's surface should be curved down beneath the center of the glacier.
   c. Continental glaciers move outward from their centers under the influence of their weight, not gravity.

9. Presently, Holland is sinking because Scandinavia is undergoing isostatic rebound.

10. A fjord is a glaciated valley that has been flooded by the ocean.

Chapter 12 Deserts

Objectives

After you study this chapter, you should be able to:

- ☐ describe the role of water in the desert
- ☐ characterize features peculiar to desert environments
- ☐ list and explain desert erosional features
- ☐ discuss the formation and types of dunes

Chapter Outline

On a seperate sheet of paper, write a single sentence that summarizes the text for each part of the following outline.

- Deserts and Wind
  - Existence
    - uneven distribution of precipitation
      - latitude
      - rain shadow deserts
      - coastal and interior deserts
  - Water and Deserts
    - desert streams
      - wash
    - desert lakes
      - playa and playa lakes
    - flooding
      - flash floods
    - pediment - erosional surface
    - bajada - depositional surface
  - Two American Deserts
    - Colorado Plateau
      - mesa
      - butte
    - Death Valley and the Great Basin
      - drainage pattern
  - Wind in Deserts
    - erosion
      - deflation
      - desert pavement
    - transport
      - abrasion

blowout

dunes

slip face

fossil dunes

cross-bedding

dune forms

Barchan

transverse

parabolic

longitudinal

loess

Checking Vocabulary

Write a definition for each of the Key Terms at the end of this chapter. Check your knowledge of these terms by answering the following questions.

1. Desert pavement is the result of
    a. deflation
    b. intense chemical weathering
    c. abrasion by windblown sand
    d. erosion by running water

2. The slip face of a particular Barchan dune is on the southeast side. The prevailing wind in this area is from the
    a. northeast
    b. northwest
    c. southeast
    d. southwest

3. The loess deposited in many parts of the Midwest
    a. had its source as glacial outwash deposits
    b. is in the form of transverse dunes
    c. is thickest and coarsest on the windward (west) side of many river valleys
    d. had its source in desert regions to the west

4. Desert and steppe regions cover about what fraction of the earth's land surface?
    a. 1/10
    b. 1/4
    c. 1/3
    d. ½

5. A wash is
    a. a desert stream channel

b. an accumulation of sand on the lee side of a dune
c. a type of loess deposit
d. a sand-free zone on the floor of a desert

6. Which one of the following is not related to wind erosion?
a. deflation
b. blowout
c. desert pavement
d. slip face

7. Dunes whose tips point into the wind are
a. Barchan dunes
c. longitudinal dunes
b. transverse dunes
d. parabolic dunes

8. Solitary dunes whose tips point downwind are
a. Barchan dunes
c. longitudinal dunes
b. transverse dunes
d. parabolic dunes

9. Long sand ridges oriented at right angles to the wind
a. Barchan dunes
c. longitudinal dunes
b. transverse dunes
d. parabolic dunes

10. Long sand ridges that are oriented more or less parallel to the prevailing wind
a. Barchan dunes
c. longitudinal dunes
b. transverse dunes
d. parabolic dunes

11. Which of the following is a wind deposit?
a. pediment
b. loess
c. Barchan
d. both b and c

12. Wind erosion is called
a. saltation
b. deflation

c. ablation
d. abrasion

13. The leeward face of a dune is not
    a. on the downwind side of the dune
    b. the side of the dune which advances forward
    c. the side of the dune where saltation is most active
    d. the steepest side of the dune

14. Depressions formed by wind erosion are called
    a. kettles
    b. cirques
    c. blowouts
    d. dunes

15. Flat-topped land surfaces in order of decreasing size are
    a. mesas, plateaus, buttes
    b. plateaus, mesas, buttes
    c. buttes, mesas, plateaus
    d. mesas, buttes, plateaus

Reviewing the Facts

Fill in the blank with the word or words which most correctly completes each statement.

1. The main factor that controls the depth of a blowout is sediment ____________.

2. As deflation lowers the surface by removing sand and silt, eventually a continuous cover of coarse particles known as _______________ remains.

3. The leeward slope of a dune, called the _______________, maintains an angle of about thirty-four degrees.

4. Dunes whose tips point into the wind are _______________ dunes.

5. Long sand ridges oriented at right angles to the wind describe _______________ dunes.

6. Long sand ridges that are oriented more or less parallel to the prevailing wind describes _______________ dunes.

7. Deposits of windblown silt are known as _______________.

8. A __________ is a dry, flat lake bed on the floor of a desert basin.

9. _________ deposits usually take the form of dunes.

10. Blowouts are shallow depressions caused by _________.

11. Desert _________ results from deflation.

12. The _______ slope of a sand dune is known as the slip face.

13. The predominant particle size in loess deposits is _________.

14. A desert receives less than _______ inches of precipitation per year.

15. Most of the earth's deserts occur at _________ north and south latitude.

Developing and Applying Concepts

1. Contrast the effect of wind on the movement of sand, silt, and clay.

2. a. How is a sand dune formed?
   b. How does a sand dune move?

3. a. How is it possible for loess to form vertical cliffs?
   b. The greatest thicknesses of loess deposits in the central Unites States are closest to the eastern edges of river flood plains. Layers of loess extend to the east and become thinner in that direction. Examination reveals that the average grain size of these deposits also decreases in the eastward direction. Explain.

4. Describe the three major factors which form deserts.

5. Why are flash floods more common in desert environments that more humid climates?

6. The Colorado River no longer flows to the Gulf of California. Give two reasons why it drys up before it reaches the ocean.

7. What are the two factors that control the size and shape of sand dunes?
8. What are the economic implications of loess deposition?

9. How has human occupation increased the desert areas of this planet?

Thinking Critically about the Environment

1. In the planning for the development of a low rainfall area, what steps would you take to prevent decertification?

2. What are the economic benefits in the development of an area which is covered by thick deposits of loess?

3. What types of aquifers could ancient wind deposits form? What would you look for to exploit these aquifers?

Answers to Questions

Checking Vocabulary

1. a, 2. a, 3. a, 4. c, 5. a, 6. d, 7. d, 8. a, 9. b, 10. c, 11. d, 12. b, 13. c, 14. c, 15. b

Reviewing the Facts

1. 2. 3. slip face, 4. parabolic, 5. Barchan, 6. longitudinal, 7. loess, 8. playa, 9. sand, 10. deflation, 11. pavement, 12. windward, 13. sand, 14. 10, 15. 30°

Developing and Applying Concepts

1. Wind carries silt in suspension. Wind may carry clay in suspension, but because of the shape of clay particles it is not often picked up by the wind. Wind can only move sand by saltation.

2. a. A dune is formed when sand begins to accumulate in the leeward space behind an obstacle. After enough accumulation, the sand begins to migrate downwind and a sand dune is formed.
b. Sand saltates up the windward side of the dune and is deposited, at the angle of repose, on the leeward side. As the sand accumulates on the leeward side, the angle of repose is exceeded and the particles slip down, moving the dune windward by a tiny fraction. By these processes a dune moves downwind.

3. a. silt particles are angular and interlock, maintaining a high angle of repose
b. the prevailing winds from the west picked up the particles and spread them eastward, with the quantities and sizes decreasing in that direction.

4. The three factors which cause deserts are:
1) latitude - global air circulation results in sinking air at 30° north and south of the equator which promotes evaporation of atmospheric moisture
2) mountains - moisture laden air, when forces up, over a mountain range loses its water which results in increased evaporation on the leeward side of the mountain range
3) relationship to ocean currents - as cool marine air encounters a land mass, it is heated and rises, encouraging more evaporation from the land's surface

5. Since deserts are characterized by rocky or tightly compacter sediment surfaces, with little or no vegetation, there is little chance for infiltration or evapotranspiration to occur, so the majority

of any precipitation moves as surface runoff.

6. The Colorado River flows through a desert and much of its water is lost by evaporation. The water which remains is diverted for use by people leaving a small enough discharge to evaporate before reaching its base level.

7. Dune formation and shape are controlled by the supply of sand and the strength of the wind.

8. Loess deposits cover the central plains and eastern Washington state in the United States which has led to the formation of fertile soils and rich farmlands.

9. Overgrazing, farming, and the gathering of firewood have all led to the increase of desert areas.

Unit IV: The Oceans

Chapter 13 Ocean Basins

Objectives

After you study this chapter, you should be able to:

- ☐ differentiate between oceanic and continental crust
- ☐ explain the techniques for studying the ocean floor
- ☐ discuss the sea-floor spreading theory
- ☐ characterize the provinces of the ocean floor
- ☐ differentiate between active and passive continental margins
- ☐ relate sea-floor features to plate motions

Chapter Outline

On a seperate sheet of paper, write a single sentence that summarizes the text for each part of the following outline.

- Earth's Oceans
  - creation
  - crust composition
  - modern oceans
    - extent
    - volume
    - size and shape
    - climatic effects
- Sea Floor
  - sampling
    - rock dredge
    - cores
    - sea-floor drilling
    - remote sensing
      - echo sounder
      - seismic profiler
      - magnetometer
      - microwave radar
- Sea-Floor Magnetism
  - magnetic patterns
    - formation

- magnetic reversals
    - polarity
- sea-floor spreading hypothesis
    - oceanic crust forms continuously
    - acquires magnetic pattern as cools
    - magnetic field reversals

Mid-Ocean Ridge
- size and extent
- rift valley
    - transform faults

Sediment and Rock of Deep Sea Floor
- three layers
    - layer 1 sediments
        - terrigenous sediment
        - pelagic sediment
        - abyssal plains
    - layer 2 and 3 basalts
        - pillow basalts
        - gabbro

Continental Margins
- types
    - passive continental margin
    - active continental margin
- passive margin characteristics
    - continental shelf
        - carbonate platforms
    - continental slope
    - continental rise
    - submarine canyons
    - abyssal fans
        - turbidity currents
- active margin characteristics
    - oceanic trench
    - shelf and slope

Oceanic Trenches and Island Arcs
- produced by subduction
- accreted terranes
    - island arc forms
    - arc migrates toward continent
    - arc is added to continent
    - low density rock resists subduction

Seamounts, Oceanic Islands, and Atolls
- process of formation

seamount
oceanic island
guyot
atoll

Checking Vocabulary

Write a definition for each of the Key Terms at the end of this chapter. Check your knowledge of these terms by answering the following questions.

1. Which one of the following is Not part of the continental margin?
   a. continental shelf
   c. continental rise
   b. continental slope
   d. continental trench

2. Atolls may be described as
   a. coral reefs paralleling a nearby coastline
   b. coral reefs surrounding a lagoon
   c. a flat-topped submerged seamount
   d. coral reefs separated from the mainland by a shallow lagoon

3. Which of the following would you not associate with turbidity currents?
   a. deposits of graded beds
   b. formation of seamounts
   c. density current
   d. sediment transport

4. Submarine canyons found on the continental slope and rise are believed to have been created
   a. by rivers during the ice age
   b. by faulting
   c. because of a plate plunging into the mantle
   d. any or all of these

5. Seamounts
   a. are a special type of oceanic trench
   b. are volcanoes that form on the ocean floor
   c. form only in the Pacific Ocean basin
   d. are submarine canyons found near Australia

6. Sediments derived primarily from the products of weathering on the continents are called
   a. lithogenous sediment
   b. biogenous sediment
   c. hydrogenous sediment

d. both a and b

7. The gently sloping submerged surface extending from the shoreline toward the deep ocean is termed
  a. continental shelf
  b. continental slope
  c. continental rise
  d. submarine canyon

8. The development of this instrument greatly enhanced our knowledge of the ocean floor.
  a. pulsar
  b. echo sounder
  c. submarine transit
  d. infrared opscanner

9. If the earth's magnetic field did not occasionally reverse polarity
  a. polar wandering curves would not exist
  b. compasses would not work
  c. the sea floor would not produce magnetic strips
  d. paleomagnetic analysis of a rock would not give the direction to the pole

10. Terrigenous sediment is sediment
  a. derived directly from land
  b. of the mid-ocean rift valley
  c. of the abyssal plain
  d. composed of mud and the remains of tiny plants and animals

11. When molten lava comes into contact with seawater, it produces a rock type called
  a. basalt dike
  b. gabbro sill
  c. pillow basalt
  d. manganese nodule

12. Features which cut across the mid-ocean ridge, formed by different rates of crustal movement are
  a. rift valleys
  b. active margins
  c. abyssal plains
  d. transform faults

13. When a magnetic reversal occurs
  a. the earth's magnetic field does not change
  b. the south magnetic pole becomes the north magnetic pole

c. the earth could lose its magnetic field
d. the magnetism no longer effects rocks

14. The topography of the ocean floor is
a. lacking any significant features
b. less varied than that of the continents
c. as varied as that of the continents
d. cannot be compared to the topography of the continents

15. The second layer of the ocean floor consists of
a. pelagic sediment
b. basalt dikes
c. pillow basalt
d. gabbro

Reviewing the Facts

Determine whether the following statements are true or false. If the statement is false, change the underlined word so it is correct.

1. Later 2 of the ocean crust contains vertical basalt dikes and gabbroic sills.

2. The last significant seismic activity along the eastern coast of north America occurred about 200 million years ago.

3. The same methods used for oil exploration have been used to obtain samples from the sea floor.

4. One of the reasons which oceanic islands tend to sink is erosion.

5. Terrigenous sediments accumulates at a rate of 2.5 millimeters every 100 years.

6. The mid-ocean ridge is characterized by a high heat flow.

7. Oceanic crust is thicker than continental crust

8. The continental shelves are of equal width in all of the earth's oceans.

9. The polarity of the earth's magnetic field has reversed two times during the earth's history.

10. Magnetic polarity stripes are arranged symmetrically on both sides of the mid-ocean ridge.

11. As distance from the mid-ocean ridge increases, sediment thickness remains the same.

12. The ocean floor has less topography as the continents.

13. The mid-ocean ridge has a width of 80,000 kilometers.

14. A passive continental margin occurs where continental and oceanic crust are firmly joined together.

15. A mountain that rises one kilometer or more above the sea floor is called an atoll.

Developing and Applying Concepts

1. a. What is the difference in the type of data obtained with an echo sounder and a seismic profiler?
   b. Which would you sue in searching for sub-sea resources?

2. Differentiate between the structure and rock composition of the mid-ocean ridge and continental mountain ranges.

3. a. Explain why someone might refer to the mid-ocean ridge as a "heat bulge".
   b. Knowing about cooling rates and density, make an inference about the relationship between the age of oceanic crust and its depth below sea level.

4. Why are earthquakes shallow along the mid-ocean ridge?

5. What is the function of a transform fault?

6. How is the reversing nature of the magnetic field used to explain the formation of the sea floor?

7. Describe the composition and origin of the two types of ocean floor sediment.

8. Compare and contrast active and passive continental margins.

9. List and describe the three factors which lead to the sinking of oceanic volcanic islands.

Thinking Critically about the Environment

1. In the exploration of the sea floor for resources, list the physical and legal problems that one might encounter.

2. What are some of the additional problems encountered when exploring for sea floor resources as opposed to looking for continental resources?

3. What are some of the advantages of using remote sensing techniques for exploration of the ocean's floors? What are the disadvantages?

Answers to Questions

Checking Vocabulary

1. d, 2. b, 3. b, 4. a, 5. b, 6. a, 7. a, 8. b, 9. c, 10. a, 11. c, 12. d, 13. b, 14. c, 15. c

Reviewing the Facts

1. Layer 3, 2. true, 3. true, 4. true, 5. pelagic, 6. true, 7. thinner, 8. unequal, 9. many, 10. true, 11. increases, 12. as much, 13. length, 14. true, 15. a seamount

Developing and Applying Concepts

1. a. The seismic profiler can obtain data on layering and sea floor strata, the echo sounder can only examine the water.
b. The seismic profiles can obtain sub-sea surface date that would be useful in resource exploration.

2. The mid-ocean ridge is composed of undeformed basalt. Continental mountain ranges are usually composed of granite and highly deformed sedimentary and metamorphic rocks.

3. a. The mid-oceanic ridge "bulges" above the rest of the ocean floor because it has a high heat flow and the newly formed basalt is not very dense.
b. The older the crust, the greater its depth below sea level.

4. Earthquakes are shallow at the mid-ocean ridge because molten rock is very close to the surface at this location and the asthenosphere in not very deep

5. Transform faults offset the mid-ocean ridge to accommodate different rates of crustal movement.

6. New basaltic crust is firming at the mid-ocean ridge, acquiring the direction of the current magnetic field as it cools. Since the magnetic field reverses occasionally, the basalt records these reversals as it expands outward from the mid-ocean ridges.

7. Terrigenous sediment is sand, silt, and clay carried from the continents by erosion and transportation by turbidity currents. Pelagic sediments are clay, carried from the continents by wind, and the carbonate or siliceous remains of microscopic plants and animals which lived at the ocean's surface.

8. Passive continental margins are not plate boundaries. They are characterized by a continental

shelf, slope, and rise - all depositional features. Active continental margins are characterized by an oceanic trench, the result of a convergent plate boundary and subduction of oceanic crust under the continent.

9. The three factors which lead to the sinking of volcanic islands are:

    1) The elimination of the mantle plume, which leads to lithospheric cooling or the motion of the plate away from the hot spot
    2) isostatic sinking as the result of increased crustal weight
    3) erosion of the exposed portions of the island.

Chapter 14 Oceans and Coastlines

Objectives

After you study this chapter, you should be able to:

- ☐ describe the chemistry of seawater
- ☐ explain the thermal structure of seawater
- ☐ characterize the features of ocean waves
- ☐ describe the motions of seawater
- ☐ discuss the interactions of waves and shorelines
- ☐ list and describe the features of emergent and submergent shorelines
- ☐ discuss the results of human intervention on shorelines

Chapter Outline

On a seperate sheet of paper, write a single sentence that summarizes the text for each part of the following outline.

- Geography of the Oceans
  - relative sizes
  - connectivity
- Seawater
  - salinity
    - dissolved ions
    - average content
    - variations
  - temperature
    - layers
      - thermocline
- Tides
  - cause
    - spring tides
    - neap tides
  - variations
- Waves and Currents
  - wave size
    - wind speed
    - wind duration
    - wind fetch
  - wave structure
    - crest

        - trough
        - wavelength
        - height
    - surface currents
        - wind driven
        - Coriolis effect
        - climatic effects
    - deep-sea currents
        - density driven
    - upwelling
        - mixing of deep and surface waters
- Coastal Weathering and Erosion
    - wave breaking
        - hydraulic action
        - abrasion
        - solution
- Sediment Transport
    - wave refraction
        - longshore current
    - beach drift
    - tidal currents
- Water's Edge
    - beaches
        - zones
            - foreshore
                - intertidal zone
            - backshore
    - reefs
        - growth
        - destruction
- Emergent and Submergent Coastlines
    - definitions
    - causes
        - tectonic
        - isostatic
        - sea level changes
            - eustatic
        - glacial
        - climatic
- Sandy and Rocky Coastlines
    - sediment-rich (sandy) coastline features
        - spit
        - baymouth bar

barrier island
lagoon
sediment-poor (rocky) coastline features
wave-cut cliff
sea arch
sea stack
fjord
estuary
Development and Pollution of Coastlines
Chesapeake Bay
Long Island
Global Warming and Sea-Level Rise
glacial melting
impact on coasts

Checking Vocabulary

Write a definition for each of the Key Terms at the end of this chapter. Check your knowledge of these terms by answering the following questions.

1. The height, length, and period of a wave depend upon
   a. the length of time the wind has blown
   b. the wind speed
   c. the fetch
   d. all of these

2. Which of the following is correct regarding a wave in the open ocean?
   a. water particles move in an almost circular path
   b. such waves are called waves of oscillation
   c. the wave form moves forward but the water particles do not advance appreciably
   d. only b and c

3. When are waves affected by the sea floor, that is, when do waves begin to "feel bottom"?
   a. when water depth is equal to one-half the wave length
   b. when water depth is equal to the wave length
   c. when water depth is twice as great as the wave length
   d. when water depth is three times as great as the wave length

4. When waves reach shallow water, they are often bent and tend to become parallel to the shore. This process is termed
   a. oscillation
   b. refraction
   c. translation
   d. reflection

5. The movement of water within the surf zone that parallels the shore is termed

a. tidal current
b. beach drift
c. rip current
d. longshore current

6. The presence of which one of the following would indicate that the land had been uplifted or that sea level had fallen?

a. an estuary
b. an elevated marine terrace
c. a sea stack
d. a tombolo

7. Which one of the following is a landform created by wave erosion?

a. spit
c. tombolo
e. breakwater
b. estuary
d. sea arch

8. One result of wave refraction is that

a. wave energy is concentrated on headlands projecting into the water
b. wave energy is concentrated in the recessed areas between headlands
c. wave energy is largely dissipated before the waves reach the shore
d. headlands are enlarged because sediment is deposited on their seaward side

9. The zigzag movement of sand grains along a beach

a. is caused by obliquely breaking waves
b. is called beach drift
c. is very unusual and seldom occurs
d. both a and b

10. A sand bar that completely crosses a bay, sealing it off from the open ocean is a

a. sea stack
b. coastal barrier
c. baymouth bar
d. tombolo

11. A ridge of sand projecting into a bay and often having a hooked end is a

a. spit
b. jetty
c. sea stack
d. groin

12. An isolated remnant of wave erosion is a
   a. spit
   b. jetty
   c. sea stack
   d. groin

13. A sand ridge connecting an island to the mainland or to another island is a
   a. jetty
   b. breakwater
   c. sea stack
   d. tombolo

14. A flat, bench-like surface cut in rock is a
   a. sea stack
   b. tombolo
   c. jetty
   d. wave-cut cliff

15. Which of the following is a tidal current?
   a. spring tide
   b. flood tide
   c. ebb tide
   d. both b and c

Reviewing the Facts

Fill in the blank with the word or words which most correctly complete each sentence.

1. The proportion of dissolved salts to pure water is termed __________.

2. Two sources for the salts in seawater are ________________ and ________________ .

3. The layer in the ocean that is characterized by a rapid temperature change is known as the ________________ and marks the transition between the warm surface layer and the deep zone of cold water.

4. Spring tides are when the variation between low tide and high tide is _________.

5. Three factors that determine the height, length, and period of ocean waves are ____________, ____________, and __________.

6. The highest part of a wave is called the ___________.

7. The depth at which wave motion becomes negligible is about __________ of the wavelength.

8. Surface ocean currents are driven primarily by the __________.

9. The ______________ effect causes ocean currents to be deflected to the right of their path of motion in the Northern Hemisphere.

10. Deep-sea currents are driven by differences in ___________.

11. The rising of cold water from deeper layers to replace warmer surface water is called ______________.

12. ____________ is the cause of most coastal erosion.

13. The refraction of waves as they approach a coast line at an angle results in a _____________.

14. The __________ and __________ are the two zones of a beach.

15. A global change in sea level, called ____________, results in the production of emergent or submergent coastlines.

Developing and Applying Concepts

1. a. Why is it said that the earth has on global ocean?
   b. What consequences does this have for pollution of the ocean?

2. a. Describe the composition of sea water.
   b. What does it mean if the concentration of a dissolved substances in a solution is 15 parts per thousand.
   c. Name two gasses which are dissolved in sea water.
   d. Why is the salinity of the Persian Gulf much higher than that of the Baltic Sea?
   e. Why can't the age of the earth be determined from the amount of salt in the ocean's waters?

3. a. list the following in the order of increasing density:
      (a) cold, salty water
      (b) warm, salty water
      (c) warm, fresh water
   b. Compare the layered structure of the ocean at the equator and in polar regions.

4. Why does a wave break?

5. Compare the motion of water as it moves onshore and returns to the ocean.

6. a. Why do tides occur almost an hour later each day?

b. What is the effect of the wind on tides?
c. Tides are higher when the Sun and the moon are on the same side of the earth than when they are on opposite sides. Why?

7. a. Compare and contrast a surface current and an upwelling.
b. Compare and contrast surface and deep ocean currents.

8. a. If samples are taken from the lower part of a reef, will the future of the reef be jeopardized? Why or why not?
b. What factors currently threaten coral reefs?

9. How could you recognize an emergent coastline on a map?

10. How does human intervention affect coastline environments?

Thinking Critically about the Environment

1. On a map, indicate the location of major ocean shipping lanes. Use a different color to indicate the location of major ocean currents. Is there a relationship between the two?

2. The oceans play a major role in maintaining the equilibrium with the concentration of carbon dioxide in the atmosphere. Make a drawing illustrating how this occurs.

3. List the information you would require about a coastal area before choosing the location for the development of residential and commercial buildings?

Answers to Questions

Checking Vocabulary

1. d, 2. d, 3. a, 4. b, 5. d, 6. b, 7. d, 8. a, 9. d, 10. c, 11. a, 12. c, 13. d, 14. c, 15. d

Reviewing the Facts

1. salinity, 2. rivers, volcanoes, 3. thermocline, 4. large, 5. wind speed, wind duration, wind fetch, 6. crest, 7. one half, 8. winds, 9. Coriolis, 10. water density, 11. upwelling, 12. abrasion, 13. longshore current, 14. foreshore, backshore, 15. eustatic

Developing and Applying Concepts

1. a. It is said the earth has one global ocean because all of the ocean basins are interconnected and water circulates between them.
b. Since water circulates between ocean basins, any pollution in one will affect them all.

2. a. Ocean water contains dissolved salts, trace elements, and dissolved gasses.
b. It means that for every 1000 parts of a solution, there are 15 parts of the dissolved substance.
c. Carbon dioxide and oxygen
d. The Baltic Sea is fed mainly by freshwater from rivers and is diluted by precipitation. The Persian Gulf area has a low rainfall, high evaporation, and few inflowing rivers.
e. Salts are incorporated into sedimentary rocks as they form.

3. a. (c), (b), (a)
b. The are three distance layers of eater at the equator:
1) a surface, warm layer about 450 meters thick,
2) a middle layer in which temperature drops rapidly as depth increases,
3) a lower layer in which the temperature is about 4° C, uniformly.
At the poles the is no layering because the water is cooled by the air, sinks, and is mixed vertically.

4. A wave breaks when the base of the wave encounters to ocean floor. Friction slows the bottom of the wave at a greater rate than the top. The crest overrides the trough and the wave becomes too steep to support itself.

5. The water moves onshore as the result of wave energy and it moves in the direction of the wave motion. The water returns to the ocean directly downslope, powered by gravity.

6. a. Tides occur about an hour later each day because the earth turns on its axis as the moon orbits the earth. This results in an additional hour of earth rotation to reach the same point in the moon's orbit.
b. Wind will enhance the tide if it is moving in the same direction, or minimize it if they are moving in opposite directions.
c. When the sun and the moon are on the same side of the earth, their combined gravitational effects will cause the highest tides possible.

7. a. In a surface current the water moves horizontally, in an upwelling the water moves vertically.
b. Both surface and deep ocean currents involve horizontal motion of water. Surface currents are powered by the winds while deep currents flow as the result of density differences.

8. a. The structural integrity of the reef may be threatened by the removal of its lower parts where corals no longer live and cannot rebuild the reef structure.
b. Pollution such as sewage and industrial wastes are damaging to reef ecosystems.

9. An emergent coastline is characterized by depositional features such as spits, barrier islands, and baymouth bars.

10. Human intervention, such as the building of groins and sea walls upset the natural movement of coastal sediments altering the patterns of erosion and deposition along coasts

Unit V: The Atmosphere

Chapter 15 The Earth's Atmosphere

Objectives

After you study this chapter, you should be able to:

- ☐ explain the differences between weather and climate
- ☐ describe the composition of the earth's's atmosphere
- ☐ discus the forms of electromagnetic radiation
- ☐ list the causes and results of the greenhouse effect
- ☐ describe the thermal structure of the atmosphere
- ☐ explain mechanisms of heat transfer between earth systems

Chapter Outline

On a seperate sheet of paper, write a single sentence that summarizes the text for each part of the following outline.

- The Earth's Atmosphere
  - weather
  - climate
- Composition of the Atmosphere
  - dry air composition
    - nitrogen
    - oxygen
    - others
  - water vapor and droplets
  - dust
- Atmospheric Pressure
  - density and weight
  - barometric pressure
    - barometers
      - mercury
      - aneroid
        - altimeter
- Solar Radiation
  - light
    - waves
      - electromagnetic radiation

            - electromagnetic spectrum
        - particles
            - photons
    - emission
        - all objects emit radiant energy
        - wavelengths
    - transmission
    - absorption
        - energy conversion
    - reflection
        - albedo
    - scattering
- Radiation Balance
    - wavelength relationships between
        - emission
        - absorption
        - reflection
        - scattering
- Temperature Changes with Elevation
    - atmospheric layers
        - troposphere
        - tropopause
        - stratosphere
        - stratopause
        - mesosphere
        - thermosphere
- Temperature Changes with Latitude and Season
    - latitude
        - intensity
    - season
        - tilt of axis
        - solstices
        - equinoxes
        - Tropics of Cancer and Capricorn
- Temperature Changes with Heat Transport and Storage
    - heat transport
        - conduction
        - convection
            - advection
    - changes of state
        - latent heat
    - heat storage
        - specific heat

Checking Vocabulary

Write a definition for each of the Key Terms at the end of this chapter. Check your knowledge of these terms by answering the following questions.

1. Which one of the following is the most abundant gas in the atmosphere?
   a. oxygen
   b. carbon dioxide
   c. hydrogen
   d. nitrogen

2. Which one of the gases listed below is more important in weather processes than the others?
   a. oxygen
   b. carbon dioxide
   c. hydrogen
   d. nitrogen

3. Fifty percent of the gases making up the atmosphere are found below
   a. 10 miles (16.2 km)
   b. 8 miles (12.8 km)
   c. 6 ½ miles (10.4 km)
   d. 3 ½ miles (5.6 km)

4. Ozone is concentrated in the
   a. troposphere
   b. mesosphere
   c. ionosphere
   d. stratosphere

5. Practically all clouds and storms occur in this layer of the atmosphere.
   a. troposphere
   b. mesosphere
   c. ionosphere
   d. stratosphere

6. The normal lapse rate or average temperature decrease in the troposphere is _______ degrees Centigrade for each kilometer of altitude increase.
   a. 3.5
   b. 6.5
   c. 1.5
   d. 1.0

7. The wavelengths emitted by the earth are
  a. longer than those emitted by the sun
  b. shorter than those emitted by the sun
  c. about the same as those emitted by the sun
  d. none of these

8. The longest wavelengths on the electromagnetic spectrum are
  a. gamma
  b. ultraviolet
  c. radio
  d. infrared

9. The storage of heat in the lower layer of the atmosphere produced by certain heat-absorbing gases is called the
  a. adiabatic effect
  d. photon effect
  b. greenhouse effect
  e. photosynthesis effect
  c. grey-body effect

10. Which one of the following latitudes best represents the location where the sun is directly overhead at noon on April 21?
  a. 0 degrees
  b. 8 degrees N
  c. 23 ½ degrees N
  d. 23 ½ degrees S

11. Which one of the following statements is true about the equinoxes?
  a. occur in June and December
  b. the sun's vertical rays are striking either the Tropics of Cancer or Capricorn
  c. days and nights are equal in length everywhere
  d. the length of daylight in the Arctic and Antarctic Circles is 24 hours

12. On this date the length of daylight gets progressively longer going south from the equator.
  a. June 21
  b. September 21
  c. March 21
  d. December 21

13. The spring equinox in the Northern Hemisphere.
  a. June 21
  b. September 21

c. March 21
d. December 21

14. New York City has its greatest length of daylight.
a. June 21
b. September 21
c. March 21
d. December 21

15. The earth receives energy from the sun in this way.
a. conduction
b. radiation
c. convection
d. all of these (a, b, & c)

Reviewing the Facts

Select one of the two italicized words in parentheses to correctly complete each of the following statements.

1. The most abundant gas in the atmosphere is *(hydrogen, nitrogen).*

2. Air is a *(mixture, compound).*

3. Annual temperature range *(increases, decreases)* with an increase in latitude.

4. *(Weather, Climate)* is the state of the atmosphere at a specific place and time.

5. Earth receives energy from the sun in the form of *(short, long)* wavelength radiation.

6. The energy absorbed by the earth is re-emitted as *(short, long)* wavelength radiation.

7. Atmospheric *(pressure, moisture)* is a measure of the weight of the atmosphere.

8. *(Oxygen, Carbon dioxide)* is a greenhouse gas.

9. Atmospheric layers are based on *(temperature, density)* changes.

10. Atmospheric temperature *(increases, decreases)* with altitude in the troposphere.

11. Ozone is concentrated in the *(mesosphere, stratosphere).*

12. Ozone filters out most of the *(ultraviolet, infrared)* radiation in sunlight.

13. Water has a very *(high, low)* specific heat.

14. Winds and ocean currents transfer heat by *(radiation, convection).*
15. In the middle and high latitudes, annual temperature range *(increases, decreases)* with an increase in continentality.

Developing and Applying Concepts

1. If there was no atmosphere how would temperatures be different on a daily basis? How would they differ on a yearly basis?

2. How have people changed the composition of the atmosphere during the past century?

3. Imagine that you are inspecting the damage produced by a tornado and notice that the glass from all the broken windows is located on the outside of the houses. Make an inference to explain this observation.

4. Compare and contrast a barometer and an altimeter.

5. a. Differentiate between the particle and wave theories of light.
   b. Relate wavelength to the frequency of electromagnetic radiation.
   c. List the following in order of increasing wavelength:
      gamma radiation
      visible light
      radio waves
      ultraviolet radiation
      infrared radiation
      x-rays
   d. list the following colors of visible light in order of increasing frequency:
      red, blue, yellow, violet, green, orange

6. Which of the following gives off the shortest wavelength radiation? Which gives off the longest? Explain.

7. a. Contrast transmission and absorption.
   b. Why are light colored clothes more comfortable in the summer than darker colored cloths?

8. What gasses in the earth's atmosphere are responsible for the greenhouse effect?

9. a. Why does the temperature of the troposphere decrease as altitude increases?
   b. Why does the temperature of the stratosphere increase as altitude increases?
   c. Why is the mesosphere cold?
   d. Why is the temperature of the thermosphere so high?

10. a. Why does the average temperature generally decrease as you travel from the equator to the poles?
b. How will the concentration and amount of solar energy received by the plants in your yard change from the day of the summer solstice to the day of the fall equinox?

11. a. Beginning with the summer solstice, describe how the direction of the tilt of the earth's axis changes with respect to the sun.
b. How would the length of a person's shadow at noon change if they were standing on the Tropic of Capricorn from the winter solstice to the spring equinox?

Thinking Critically about the Environment

1. There is a difference of opinions as to whether the burning of tropical rainforests affects the composition of the earth's atmosphere. Choose a side of this debate to support.

2. Develop a plan to assess the impact of trees on the air quality of urban environments.

3. How has the concentration of carbon dioxide in the atmosphere changed during the past 150 years? List the possible impacts of this change.

4. How could human activities deplete the ozone layer in the stratosphere. Research how this could have happened and the steps being taken to solve this problem.

Answers to Questions

Checking Vocabulary

1. d, 2. b, 3. d, 4. d, 5. a, 6. b, 7. a, 8. c, 9. b, 10. b, 11. c, 12. d, 13. c, 14. a, 15. b

Reviewing the Facts

1. nitrogen, 2. mixture, 3. decreases, 4. weather, 5. short, 6. long, 7. pressure, 8. carbon dioxide, 9. temperature 10. decreases, 11. stratosphere, 12. ultraviolet, 13. high, 14. convection, 15. increases

Developing and Applying Concepts

1. If there was no atmosphere on earth , the temperatures would be much higher during the day and much lower at night. The yearly range of temperatures would also be greater.

2. Humans have added large amounts of sulfur dioxide, nitric oxide, soot, smoke, and

dust to the atmosphere. Human actions have also increased the amount of carbon dioxide and methane in the atmosphere.

3. The air pressure on the inside of the houses must have been greater than the air pressure of the tornado causing the air to move outward, carrying the glass with it.

4. A barometer and an altimeter both measure changes in air pressure. A barometer is calibrated to measure air pressure, while an altimeter is calibrated to measure altitude.

5. a. According to the particle theory, light consists of packets of energy called photons. The wave theory states that electromagnetic radiation is a continuous form of energy.
   b. the longer the wavelength, the lower the frequency
   c. gamma radiation, x-rays, ultraviolet radiation, visible light, infrared radiation, radio waves
   d. red, orange, yellow, green, blue, violet

6. Molten lava gives off the shortest wavelengths because it is the hottest. A glacier would give off the longest wavelength radiation because it is the coolest.

7. a. Transmitted light is passed through an object without loss of energy. Light that is absorbed is converted to another form of energy.
   b. Light colored clothing reflect more of the visible light, darker colored clothe absorb the visible light and convert it to heat.

8. Water vapor, carbon dioxide, and methane are a few of the greenhouse gasses.

9. a. The source of the heat for the troposphere is the infrared radiation given off by the earth's surface. As altitude increase, the distance from the heat source increases.
   b. The source of heat fro the stratosphere is the ultraviolet radiation absorbed by the ozone layer.
   c. Little electromagnetic radiation is absorbed in the mesosphere.
   d. Gamma radiation and x-rays ionize the atmospheric gasses in the thermosphere.

10. a. The concentration of solar radiation decreases from the equator to the poles.
    b. The amount and concentration of solar radiation the plants receive will decrease.

11. a. On the day of the summer solstice the north pole of the earth's axis is pointed directly toward the sun. On the day of the winter solstice it is pointed directly away from the sun. On the spring and fall equinoxes, the axis is not tilted either toward or away from the sun.

b. The shadow would lengthen.

Chapter 16 Weather

Objectives

After you study this chapter, you should be able to:

- ☐ describe how humidity is measured
- ☐ explain how condensation occurs
- ☐ list and describe the three basic types of clouds
- ☐ discuss how winds form
- ☐ explain vertical and horizontal air motions
- ☐ describe frontal weather
- ☐ describe long- and short-term wether disturbances

Chapter Outline

On a seperate sheet of paper, write a single sentence that summarizes the text for each part of the following outline.

- Moisture in Air
    - humidity
        - absolute humidity
        - relative humidity
    - saturation
        - dew point
        - supersaturation
        - supercooling
- Cooling and Condensation
    - mechanisms
        - radiation cooling
        - contact cooling
        - rising
            - adiabatic temperature change
            - adiabatic lapse rate
                - wet adiabatic lapse rate
                - dry adiabatic lapse rate
- Clouds
    - rising mechanisms
        - convection
        - frontal wedging
        - orographic lifting
    - cloud formation

        - normal lapse rate
    - cloud types
        - cirrus
        - stratus
        - cumulus
        - combined forms
            - stratocumulus
            - cumulonimbus
            - nimbostratus
            - altostratus
    - precipitation
        - rain
            - collide and coalesce
            - ice crystal
        - snow
        - sleet
        - glaze
        - hail
            - cumulonimbus clouds
- Fog
    - advection fog
    - radiation fog
    - evaporation fog
    - upslope fog
- Pressure and Wind
    - wind
        - horizontal air flow
    - pressure gradient
    - Coriolis effect
    - friction
        - jet stream
    - cyclones and anticyclones
    - pressure and weather
- Fronts and Frontal Weather
    - air masses
        - polar
        - tropical
        - Maritime
        - Continental
    - fronts
        - warm front
        - cold front
        - occluded front

- stationary front
- mid-latitude cyclone
- Mountains, Oceans, Lakes, and Weather
  - rain shadow desert
  - land and sea breezes
  - monsoon
- Thunderstorms
  - size and duration
  - rising air
    - wind convergence
    - convection
    - orographic rise
    - frontal rising
  - stages of development
    - wind shear
- Tornadoes and Tropical Cyclones
  - tornado
    - description
    - duration
    - damage
  - tropical cyclone
    - names
      - hurricane
      - typhoon
      - cyclone
    - structure
    - damage

Checking Vocabulary

Write a definition for each of the Key Terms at the end of this chapter. Check your knowledge of these terms by answering the following questions.

1. The cloud form that consists of globular cloud masses that take on a billowy or cotton ball structure is called
   a. cumulus
   b. cirrus
   c. stratus
   d. alto

2. This term is used to describe clouds found in the middle height range
   a. cumulus
   b. cirrus
   c. stratus

d. alto

3. High clouds that form delicate veil-like patches or extended wispy fibers are termed
   a. cumulus
   b. cirrus
   c. stratus
   d. alto
   e. nimbo

4. Thunder and lightning are associated with these clouds
   a. cirrostratus
   b. altostratus
   c. nimbostratus
   d. cumulonimbus

5. This fog forms in valleys at night
   a. advection fog
   b. precipitation fog
   c. radiation fog
   d. steam fog

6. When warm moist air moves over a cold surface, ______________ fog may result
   a. advection
   b. upslope
   c. steam
   d. radiation

7. We are likely to have our highest relative humidity
   a. during mid-afternoon
   b. at noon
   c. about midnight
   d. about sunrise

8. The most important process of cloud formation in the atmosphere is
   a. cooling by compression of air
   b. cooling by release of latent heat of vaporization
   c. cooling by expansion of air
   d. radiation cooling

9. The force that generates wind is
   a. Coriolis force
   b. friction
   c. gravity

d. pressure gradient force

10. Lines on a weather map connecting places of equal air pressure are termed
   a. isovectors
   b. isogrids
   c. isopressure
   d. isobars

11. In the northern hemisphere, the winds associated with a low pressure system (cyclone) blow
   a. clockwise toward the center
   b. counterclockwise toward the center
   c. clockwise outward from the center
   d. counterclockwise outward from the center

12. If stormy weather is approaching, the pressure tendency would probably be
   a. steady
   b. falling
   c. rising
   d. pressure tendency has nothing to do with forecasting good or bad weather

13. If an observer sees cirrus clouds, followed later by cirrostratus, and then altostratus, he or she is witnessing the approach of a ______________ front.
   a. warm
   b. cold
   c. either a or b
   d. none of these

14. When an active cold front overtakes a warm front
   a. the fronts cancel one another out
   b. cloud formation ceases
   c. an occluded front forms
   d. a stationary front is created

15. Which of the following statements is not true of tornadoes?
   a. occur most frequently in the spring of the year
   b. usually occur along the warm front of a mid-latitude cyclone
   c. most common in the Midwest and Great Plains
   d. associated with cumulonimbus clouds

Reviewing the Facts

Fill in the blank with the word or words which most correctly completes each statement.

1. The term used to describe clouds that form sheets or layers that cover much of the sky is

_______________.

2. The term used to indicate a cloud found in the middle height range is _______________.

3. High clouds that form delicate veil-like patches or wispy fibers are called _______________.

4. Thunder and lightning are normally associated with _______________ clouds.

5. Name the small billowy clouds associated with fair weather. _______________

6. What type of fog forms in valleys at night? _______________

7. When warm moist air moves over a cold surface, _______________ fog may result.

8. When supercooled raindrops freeze on contact with solid objects, _______________ forms.

9. The change of state from a gas to a liquid is called _______________.

10. The term _______________ is used to describe the conversion of a solid directly to a gas, without passing through the liquid state.

11. The process of converting a liquid to a vapor is termed ____________.

12. A measurement which indicates how near the air is to saturation is called _______________.

13. _______________ temperature changes are those which do not involve the addition or subtraction of heat.

14. _______________ wedging occurs when warm air is forced over cooler air.

15. Deserts which are cut off from moisture by mountain ranges are termed _______________ deserts.

16. The _______________ is the temperature to which a parcel of air would need to be cooled in order to reach saturation.

Developing and Applying Concepts

1. a. Air A contains 8.0 g/m^3 of water vapor and can hold 11.0 g/m^3 at its current temperature. Air B contains 7.0 g/m^3 and can hold 9.0 g/m^3 at its current temperature. Which air has the greatest relative humidity?
   b. Which air described above has the greatest absolute humidity?
   c. Which air described above is the least saturated?

d. What will happen to the relative humidity of Air A if its temperature decreases?
e. Air C contains 10.0 g/m3 of water vapor and has a relative humidity of 80%. What is the maximum amount of water vapor it can hold at its current temperature?

2. a. List three basic requirements for cloud formation.
b. What factor determines the height at which cloud formation begins?

3. a. If dry air that has a temperature of 50° C rises 4000 meters, what will its temperature become?
b. If wet air that is 75° C rises 3000 meters, what will its temperature become?

4. Compare and contrast cumulonimbus and nimbostratus clouds.

5. Would you expect fog to form under the following circumstances? Explain.
a. warm, moist air is situated over a large lake in the summertime
b. cold, dry air is forced up a mountainside
c. warm, moist air is cooled when it moves inland from the sea

6. a. What type of air mass will form over a subtropical desert, high or low pressure?
b. What type of air mass will form over a continental glacier, high or low pressure?
c. If the two air masses meet, in what direction will the wind blow?
d. What will determine the strength of the wind?

7. How does the circulation pattern of a cyclone in the Northern Hemisphere compare to the circulation pattern of a cyclone in the Southern Hemisphere?

8. a. Compare and contrast a warn front and a cold front.
b. How does an occluded front form?

9. Identify the condition which could lead to the formation of a thunderstorm in each of the following areas:
a. the Great Plains of the United States
b. the Rock Mountains
c. central Florida

10.Why do hurricanes dissipate after they move over land?

Thinking Critically about the Environment

1. Would an increase in global temperatures have an effect on the local weather? Explain.

2. Plot the tracks of the hurricanes which occurred during the past few years. How could you use this data to predict the timings and locations of future storms?

3. During the design phase of a new community, what measures could be taken to minimize the threats posed by large-scale storms?

Answers to Questions

Checking Vocabulary

1. a, 2. d, 3. b, 4. d, 5. c, 6. a, 7. d, 8. c, 9. d, 10. d, 11. b, 12. b, 13. a, 14. c, 15. b

Reviewing the Facts

1. stratus, 2. alto, 3. cirrus, 4. cumulonimbus, 5. cumulus, 6. radiation, 7. evaporation, 8. glaze, 9. condensation, 10. sublimation, 11. evaporation, 12. humidity, 13. adiabatic, 14. frontal, 15. rain shadow, 16. dew point

Developing and Applying Concepts

1. a. Air B has the greatest relative humidity. Air B's relative humidity is 78%, Air A's relative humidity is 73%.
Air A has a greater absolute humidity because it contains more water per unit volume.
c. Air A is least saturated.
d. If the temperature of Air A decreases, its relative humidity will go up. Its absolute humidity will stay the same.
e. 12.5 g/m3, divide 10 g/m3 by 0.80

2. a. Air containing water vapor, a means for cooling the air, and particles for the water vapor to condense onto are the three basic requirements for the formation of clouds.
b. The dew point temperature determines the altitude at which a cloud will form.

3. a. The air cools 10° C for every 1000 meters it rises, so its temperature will decrease by 40° C to become 10° C.
b. Wet air cools 5° C for every 1000 meters it rises, so its temperature will become 60° C.

4. Both cumulonimbus and nimbostratus clouds are rain clouds. Cumulonimbus clouds have more vertical development than the horizontally layered nimbostratus clouds.

5. a. No. Evaporation fog usually occurs in the late fall or winter when the temperatures are cooler
b. No. If the air doesn't contain moisture, up slope fog cannot form
c. Yes. Advection fog will form if warm, moist air is cooled when it moves inland

6. a. A low pressure air mass will form over a desert because it will contain warm air
b. A high pressure air mass will form over a continental glacier because it will contain cold air

c. The wind will blow from high to low pressure
d. the pressure gradient will determine the strength of the wild

7. Cyclones in the Northern Hemisphere have a counterclockwise circulation, while those in the Southern Hemisphere have a clockwise circulation pattern.

8. a. Warm and cold fronts are distinguished by their direction of movement. A warm front is where warmer air is moving over cooler air, a cold front is where cooler air pushes into warmer air.
b. An occluded front forms when warmer air is trapped between two cooler air masses.

9. a. Convection could cause thunderstorms on the Great Plains as the sunlight would warm the ground and generate cells of rising, moist air
b. Orographic lifting of moist air over the mountains could generate thunderstorms
c. Sea breezes could converge and generate thunderstorms. As the sun heats the land surface, the warm air rises and moist air is drawn inland.

10. Hurricanes dissipate when they move over land because evaporation cannot occur to provide the energy to fuel the storm. Once the source of water vapor is eliminated, condensation will no longer occur and the storm cycle will be broken.

Chapter 17 Climate

Objectives

After you study this chapter, you should be able to:

- ☐ explain how solar radiation influences climate
- ☐ list other factors which modify climate
- ☐ describe the three-cell model of atmospheric circulation
- ☐ characterize climate zones

Chapter Outline

On a seperate sheet of paper, write a single sentence that summarizes the text for each part of the following outline.

- Major Factors that Control Climate
    - latitude
    - winds
    - oceans
    - altitude
    - albedo
    - combinations
- Global Winds and Climate
    - global circulation - three cell model
        - doldrums
        - horse latitudes
        - trade winds
        - prevailing westerlies
        - polar easterlies
        - polar front
        - jet streams
            - subtropical
            - polar
- Ocean Currents and Climate
    - surface currents
    - deep currents
    - vertical mixing
- Climate Zones
    - Koeppen climate classification
        - A - humid tropical climates
            - abundant rainfall
                - rainforests

seasonal rainfall
tropical monsoon
tropical savanna
B - dry climates
C - mid-latitude climates (mild winter)
humid subtropical
Mediterranean
marine west coast
temperate rain forest
D - humid mid-latitude (severe winter)
E - polar climates
Urban Climates
heat island effect

Checking Vocabulary

Write a definition for each of the Key Terms at the end of this chapter. Check your knowledge of these terms by answering the following questions.

1. A narrow band of fast-moving high altitude air which flows at 30° latitude is the
   a. subtropical jet
   b. doldrums
   c. polar jet
   d. polar front

2. The type of climate which extends poleward form the Mediterranean zones to 65° is
   a. humid subtropical
   b. marine west coast
   c. tropical savanna
   d. taiga

3. A high pressure belt circling the globe at 30° is called
   a. polar front
   b. horse latitudes
   c. trade winds
   d. subtropical jet

4. A type of climate with distinct dry and wet seasons that supports the growth of rain forests is
   a. tropical monsoon
   b. tropical savanna
   c. tropical rainforest
   d. humid subtropical

5. The number of cells in the currently accepted model of the atmosphere is

a. one
b. two
c. three
d. four

6. A community of plants living in a large geographic area characterized by a particular climate is called a
   a. rainforest
   b. biome
   c. grassland
   d. taiga

7. The difference in temperature between cities and outlying areas is called the
   a. Coriolis effect
   b. Greenhouse effect
   c. Milankovitch effect
   d. urban heat island effect

8. The type of climate which occurs on the west coasts of all continents between latitudes of 30° and 40°, characterized by rainy winters, dry summers, and moderate temperatures is
   a. humid subtropical
   b. Mediterranean
   c. marine west coast
   d. taiga

9. The biome typical of the subarctic portion of climate zone D is
   a. humid subtropical
   b. tropical savanna
   c. taiga
   d. tundra

10. The global winds between 30° and 60° latitude are called
   a. doldrums
   b. prevailing westerlies
   c. trade winds
   d. polar easterlies

11. The global winds between 30° and 0° latitude are called
   a. doldrums
   b. prevailing westerlies
   c. trade winds
   d. polar easterlies

12. The type of rainforests common along the northwest coast of North America, from Oregon to Alaska are called

a. tropical
b. humid
c. temperate
d. marine

13. Ocean currents modify climates by

a. transporting heat
b. reflecting solar radiation
c. wave refraction
d. altering wind patterns

14. The biome characteristic of polar regions is

a. taiga
b. tundra
c. polar front
d. polar jet

15. A low-pressure boundary zone located at 60° latitude is called

a. polar easterlies
b. doldrums
c. polar front
d. polar jet

Reviewing the Facts

Determine whether the following statements are true or false. If the statement is false change the underlined word so that it is correct.

1. High latitudes experience greater variations in solar energy than lower latitudes.

2. Oceans effect climate in three ways.

3. Albedo is a measure of the amount of sunlight absorbed.

4. George Hadley proposed a three-cell model for atmospheric circulation.

5. The jet stream is the result of the earth's revolution.

6. Near the equator, steady winds are rare.

7. The doldrums occur at about 30° north and south latitude.

8. The wind belt covering most of the contiguous United States is called the trade winds.

9. The subtropical jet stream flows between the trade winds and the westerlies.

10. The Koeppen Climate Classification defines three principle climatic groups.

11. Tropical monsoon and savanna climates are characterized as having small seasonal variations in rainfall.

12. The earth's largest deserts lie along the 30° latitude high-pressure zones.

13. Mediterranean climates is characterized by rainy summers and dry winters.

14. Temperate rainforests are found in marine east coast climates.

15. Urban environments are generally warmer than surrounding rural areas.

Developing and Applying Concepts

1. Draw a diagram of the earth's global wind circulation. Label the latitudes, the common names of seven wind and pressure zones, and four jet streams. Use arrows to show the wind directions.

2. Why isn't it possible to predict climate on the basis of geographic position?

3. a. What two factors are the most important in defining the climate of an area?
b. What types of climates support the growth of extensive grasslands and rainforests?
c. What two types of climates support only sparse or low-lying vegetation?

4. a. Group the following environments according to the type of vegetation they contain:
humid subtropical
semiarid
tropical rainforest
marine west coast
wet, humid mid-latitude with severe winters
polar
tropical monsoon
desert
dry, humid mid-latitude with severe winters
tropical savanna
Mediterranean
subarctic
b. Group the climates into the following vegetation categories: rainforests, grasslands, shrubs with scattered trees, sparse vegetation, forests of coniferous or coniferous and

deciduous trees.

5. a. Explain why it generally rains more over cities than rural areas.
   b. How much more cloud cover do cities experience when compared to rural areas?

6. List and describe the impact of ocean currents on climate.

7. Define a biome and explain what it can tell us about regional climates.

Thinking Critically about the Environment

1. Construct a map showing the earth's five major climatic zones. How does the distribution of climatic areas affect human population distribution?

2. In planning for the development of alternative forms of energy production, how would you use the global atmospheric circulation model? What modifications would you incorporate into this model?

3. Research the theories that are attempting to explain how vertical ocean mixing affects the climate.

4. If you were to design a city, how would you go about minimizing the urban heat island effect?

Answers to Questions

Checking Vocabulary

1. a, 2. b, 3. b, 4. a, 5. c, 6. b, 7. d, 8. b, 9. c, 10. b, 11. a, 12. c, 13. a, 14. b, 15. c

Reviewing the Facts

1. true, 2. two, 3. reflected, 4. two-cell, 5. rotation, 6. true, 7. horse latitudes, 8. westerlies, 9. true, 10. five, 11. large, 12. true, 13. dry, rainy, 14. west coast, 15. true

Developing and Applying Concepts

1. a. draw a circle with a horizontal line dividing it in the center and label this line "equator"
   b. divide the upper and lower halves into three, labeling the lines 30° and 60° north and south latitude respectively
   c. label "doldrums" along the equator, "horse latitudes" at 30°, and "polar front" at 60° in both the northern and southern Hemispheres
   d. draw arrows from the north east to the south west between 0° and 30° north latitudes and label them "trade winds"
   e. draw arrows from the south east to the north west between 0° and 30° south latitudes

and label them "trade winds"
f. draw arrows from the south west to the north east between 30° and 60° north latitude and label this section "prevailing westerlies"
g. draw arrows from the north west to the south east between 30° and 60° south latitude and label this section "prevailing westerlies"
h. draw arrows between 60° and 90° north latitude similar to those between 0° and 30° north latitude and label them "polar easterlies"
I. draw arrows between 60° and 90° south latitude similar to those between 0° and 30° south latitude and label them "polar easterlies"
j. label the polar jet streams at 60° north and south latitudes and the subtropical jet streams at 30° north and south latitudes

2. Many more factors affect climate other than latitude, longitude and distance from the ocean.

3. a. temperature and precipitation
b. grasslands:
tropical savanna
tropical monsoon climates
marine west coast
rainforests:
tropical climates with high rainfall
tropical monsoon climates
marine west coast
c. arctic and subarctic

4. Rainforests
tropical rainforests
tropical monsoon
marine west coast
Grasslands
tropical savanna
semiarid
dry, humid mid-latitude with severe winters
Shrubs with scattered trees
Mediterranean
Forests of Coniferous Trees and Coniferous and Deciduous Trees
subarctic
wet, humid mid-latitude with severe winters
humid subtropical
Sparse vegetation
deserts
polar

5. a. Local low pressure zones form over cities because the buildings and other stone structures retain heat. Dust particles form condensation nuclei for water vapor. The increases atmospheric friction causes weather systems to linger longer than in rural areas.
b. Cities experience 5% to 10% more cloud cover than rural areas.

6. Ocean currents transport heat, altering regional temperatures. They also affect rainfall, noting the abundant rain which falls on the north west United States as the result of the warm Pacific current.

7. A biome is a community of plants living in a large geographic area characterized by a particular climate.

Unit VI: Human Interactions with Earth Systems

Chapter 18 Climate Change

Objectives

After you study this chapter, you should be able to:

- ☐ differentiate between feedback mechanisms and threshold effects
- ☐ list the techniques used to measure climatic changes
- ☐ describe the movement of carbon through earth systems
- ☐ characterize human impacts on climate
- ☐ list natural mechanisms for climate change
- ☐ describe astronomical causes for climate change

Chapter Outline

On a seperate sheet of paper, write a single sentence that summarizes the text for each part of the following outline.

- Climate Change in Earth History
    - composition of primordial atmosphere
    - ice core data
    - threshold and feedback mechanisms
        - glacial melting
        - earth's orbit
        - albedo changes
- Measuring Climate Change
    - historical records
    - tree rings
    - pollen
    - isotopes in glacial ice
    - glacier features
    - plankton and isotopes
    - rock and fossil record
- Carbon Cycle and Climate
    - carbon reservoirs
        - atmosphere
        - biosphere
        - hydrosphere
        - lithosphere

- carbonate rocks
- fossil fuels
- sea-floor sediments
- mantle

Humans, Carbon Cycle, and Climate
- greenhouse gasses
  - carbon dioxide
  - methane
  - CFC's (chloroflurocarbons)
  - nitrogen oxides
- greenhouse warming effects
  - temperature
  - precipitation
  - soil moisture
- possible mitigations/solutions

Changes in Ocean Circulation
- ocean/atmosphere interactions

Tectonics and Climate Change
- continental positions
- mountain ranges
- volcanic activity
- sea-level changes
- weathering

Astronomical Causes of Climate Change
- solar radiation
  - solar output
- bolide impacts
- dust and gas
- solar evolution

Checking Vocabulary

Write a definition for each of the Key Terms at the end of this chapter. Check your knowledge of these terms by answering the following questions.

1. The early earth was warmer than present day climates because
   a. the sun produced more energy
   b. the earth's orbit was closer to the sun
   c. the atmosphere contained a higher concentration of greenhouse gasses
   d. the global albedo was lower

2. Tree rings can be used to record climatic variations because
   a. ring width increases during warm, wet years
   b. ring numbers increase during warm, wet years

c. ring width decreases during warm, wet years
d. ring numbers decrease during warm, wet years

3. Oxygen isotope ratios in water change as the result of
   a. changes in atmospheric concentrations
   b. changes in water temperatures
   c. changes in ice compaction rates
   d. changes in water depths

4. Carbon exists in the atmosphere predominantly as
   a. carbon dioxide and methane
   b. carbonates
   c. hydrocarbons
   d. chloroflurocarbons

5. Carbon may be moved from the atmosphere to the crust by
   a. organic decay
   b. respiration
   c. combustion
   d. production of fossil fuels

6. The amount of carbon dioxide dissolved in ocean waters
   a. is constant
   b. varies with atmospheric composition
   c. depends on temperature
   d. varies with atmospheric pressure

7. Most of the earth's carbon may be found in the
   a. atmosphere
   b. biosphere
   c. hydrosphere
   d. lithosphere

8. Carbon may be stored in sea-floor sediments as
   a. organic skeletons
   b. carbon dioxide
   c. methane hydrate
   d. fossil fuel

9. Carbon is released from the deep mantle by
   a. subducting tectonic plates
   b. volcanic eruptions
   c. recovering fossil fuels

d. organic respiration

10. Which of the following is not considered a greenhouse gas?
    a. oxygen
    b. carbon dioxide
    c. nitrogen oxide
    d. chloroflurocarbons

11. Large amounts of which greenhouse gas are emitted by agricultural processes?
    a. carbon dioxide
    b. nitrogen oxide
    c. methane
    d. chloroflurocarbon

12. Which is the most abundant industrial greenhouse gas?
    a. carbon dioxide
    b. nitrogen oxide
    c. methane
    d. chloroflurocarbon

13. How could an increase in greenhouse gas concentration cause global cooling?
    a. prevention of incoming solar radiation
    b. increase in ocean water evaporation
    c. decrease in global glacial cover
    d. increase in plant growth rates

14. Natural causes of climate change include
    a. plate tectonic activities
    b. volcanic eruptions
    c. sea level fluctuations
    d. all of the above

15. Extraterrestrial caused of climate change include
    a. fluctuations in solar output
    b. variations in the earth's orbit
    c. bolide impacts
    d. all of the above

Reviewing the Facts

Select one of the two italicized words in parentheses to correctly complete each of the following statements.

1. Climate change in *(unique, not unique)* to the twentieth century.

2. Earth's *(earliest, present)* contains very high concentrations of carbon dioxide and water vapor.

3. A feedback mechanism occurs when a *(large, small)* initial perturbation affects another component, amplifying the original effect.

4. Climate change may be measured using *(oxygen, nitrogen)* isotope ratios in glacial ice.

5. The formation of *(carbonate, igneous)* rocks can be used to calculate past atmospheric and temperature conditions.

6. The element *(hydrogen, carbon)* is stored in sea water in solution.

7. *(Nitrogen, Carbon)* in a major component in fossil fuels.

8. When organic material falls to the sea floor, it decomposes to release *(methane, helium).*

9. Carbon can be stored in the *(crust, deep mantle)* by subduction of oceanic crust.

10. During the past 100 years, human have become a *(major, minor)* part of the carbon cycle.

11. Larger amounts of methane are released by *(industrial, agricultural)* processes.

12. It is predicted that flooding *(will, will not)* become more common as global temperatures rise.

13. The United States *(has, has not)* ratified an international treaty to control the emission of greenhouse gasses.

14. One way in which the earth's *(surficial, tectonic)* processes have regulated climate change is through the position of the continents.

15. Impacts of extraterrestrial objects *(have, have not)* changed the earth's climate in the past.

Developing and Applying Concepts

1. List seven methods for measuring climate during the earth's past.

2. Explain how oxygen isotope ratios indicate past temperatures.

3. List four natural objects which can release carbon dioxide gas into the atmosphere. What changes in each cause an increase in the carbon dioxide content of the atmosphere?

4. List and explain four ways in which humans increase the atmospheric content of greenhouse gasses.

5. Some scientists argue that an increase in the concentration of greenhouse gasses in the atmosphere could eventually cause an decrease in global temperatures. Explain how this could possibly occur.

6. List and explain four ways in which the earth's tectonic processes may alter global climates.

7. a. What uncertainties exist in predicting future trends in the earth's climate?
b. How would a significant increase in global temperatures affect the earth?
c. What are three ways to decrease the amount of carbon dioxide released in the atmosphere by burning fossil fuels?

Thinking Critically about the Environment

1. Construct a diagram illustrating the carbon cycle. Use arrows to show the directions of flow between each of the earth's spheres.

2. Make a list of the ways in which your daily activities could impact the flow of carbon in the carbon cycle. What are some measures you could take to minimize your impacts?

3. Examine the treaty that resulted from the Kyoto conference on global warming. Should the United States ratify this treaty? Build an argument supporting your stand.

4. Make a list of mechanisms of climate change. Separate those on which human's have control, human's may possibly effect, and those beyond human impact.

Answers to Questions

Checking Vocabulary

1. c, 2. a, 3. b, 4. a, 5. d, 6. c, 7. d, 8. c, 9. b, 10. a, 11. c, 12. a, 13. b, 14. d, 15. d

Reviewing the Facts

1. unique, 2. earliest, 3. small, 4. oxygen, 5. carbonate, 6. carbon, 7. carbon, 8. methane, 9. deep mantle, 10. major, 11. agricultural, 12. will, 13. has not, 14. tectonic, 15. have

Developing and Applying Concepts

1. Seven methods for measuring climate change during the earth's past are:
historical records
tree ring thicknesses
plant pollen
oxygen isotope ratios in glacial ice
glacial evidence

plankton and oxygen isotopes in ocean sediments
rock and fossil record

2. Of the two most common oxygen isotopes, during warmer periods more ^{18}O is found in water vapor then ^{16}O. This ratio is preserved in glacial ice and in the shells and skeletons of organisms which lived during there times.

3. Sea water - When sea water is heated, dissolved carbon dioxide escapes.
Limestone - CO_2 is released as a gas when limestone dissolves.
Living tissue - CO_2 is released when plants and animals die, and decay.
Rocks - Volcanic eruptions release carbon dioxide when tectonic plates move.

4. The combustion of fossil fuels releases carbon dioxide.
Logging and the clearing of rainforests eliminates plants which remove CO2 from the atmosphere.
Methane is released during industrial and agricultural processes.
Chloroflurocarbons, which were used as propellants and refrigerants, when released into the atmosphere, decrease the stratospheric ozone content.

5. An increase in the concentration of greenhouse gasses in the atmosphere could eventually cause an increase in global temperatures as the result of oceanic warming, causing more evaporation, leading to an increase in cloud cover, which would cause more of the solar radiation to be reflected, thus resulting in overall cooling.

6. Four of the ways in which tectonic processes can modify climate are:
a. positions of the contents relative to latitude and the directions of ocean currents.
b. mountain ranges alter the flow paths and speeds of global winds
c. volcanic activity adds particulates and carbon dioxide to the atmosphere
d. changes in sea level caused by tectonic activities alters ocean-atmosphere interactions.

7. a. The reliability of temperature data collected in the field is disputed. Most weather stations are neat cities and subject to the urban heat island effect. There are fewer data collecting stations in the Southern Hemisphere and almost none in the oceans. The causes of any trend are many and not easily related to their effects.
b. Rainfall patterns could change resulting in desert areas receiving much more rainfall than agricultural areas. Natural ecosystems would be disrupted by the extinction of species. Sea level would rise as the result of the melting of glacial ice.
c. (1) burn less fuel
(2) increase energy efficiency
(3) increase the use of alternative energy resources

Chapter 19 Air Pollution

Objectives

After you study this chapter, you should be able to:

- ☐ list sources of air pollution
- ☐ explain the processes which may produce an inversion
- ☐ describe the reactions which cause acid precipitation and smog
- ☐ discuss the impact of chlorofluorocarbons on the ozone layer
- ☐ describe why air pollution has not been controlled

Chapter Outline

On a seperate sheet of paper, write a single sentence that summarizes the text for each part of the following outline.

- History of Air Pollution
    - individual sources
    - general air quality
- Sources of Air Pollution
    - burning
    - volatiles
    - particulates and aerosols
    - secondary pollutants
- Meteorology of Air Pollution
    - turbulence
    - inversion
- Air Pollution and Health
    - exposure
    - concentration
- Acid Rain
    - pH scale
    - contaminates
    - consequences
    - particulates
- Smog and Ozone in the Troposphere
    - smog components
    - consequences
    - ozone
- Air Pollution from Toxic Volatiles
    - dioxin
    - sources

Air Pollution from Aerosols
- chemistry and toxicity
- albedo

Depletion of the Ozone Layer
- cause
- result

Air Pollution: Looking Back and Forward
- economic effects
- environmental balance

Checking Vocabulary

Write a definition for each of the Key Terms at the end of this chapter. Check your knowledge of these terms by answering the following questions.

1. Which one of the following is not a primary pollutant?
    a. sulfuric acid
    c. nitrogen oxide
    b. carbon monoxide
    d. particulate matter

2. A value of ______ on the pH scale indicates the solution is neutral (neither acid or alkaline).
    a. 0
    b. 5
    c. 7
    d. 10

3. A colorless, odorless, toxic gas which interferes with the body's ability to absorb oxygen is
    a. nitrogen dioxide
    b. ozone
    c. carbon monoxide
    d. nitrogen

4. A compound which evaporates readily is called a
    a. volatile
    b. particle
    c. hydrocarbon
    d. secondary pollutant

5. Ozone is
    a. a primary air pollutant
    b. a secondary air pollutant
    c. nontoxic
    d. a volatile

6. Pollutants may
   a. "seed" clouds and thus cause increased rainfall over cities
   b. cause a reduction in solar radiation received at a city
   c. add to the heat island effect
   d. all of the above

7. Carbon dioxide is
   a. toxic
   b. a combustion product
   c. carcinogenic
   d. a key component of smog

8. Acid precipitation is
   a. a primary air pollutant
   b. a secondary air pollutant
   c. nontoxic
   d. a volatile

9. Atmospheric inversions
   a. are the same thing as temperature inversions
   b. are usual daytime occurrences in most places
   c. rarely occur in Los Angles
   d. are most destructive over the oceans

10. Ozone is beneficial to organisms when
   a. used as an aerosol propellant
   b. concentrated in the stratosphere
   c. incorporated within a chlorofluorocarbon
   d. eliminated from the biosphere

11. The energy to convert automobile exhaust to smog comes from
   a. burning fossil fuels
   b. the production of ozone
   c. acid precipitation
   d. solar radiation

12. Any pollutant larger than a molecule is called
   a. soot
   b. a volatile
   c. fly ash
   d. a aerosol

13. The Clean Air Act

a. sets national air quality standards
b. insures that we will not have to breathe unhealthy air
c. will eliminate air pollution by 2002
d. is vigorously enforced

14. The primary source of sulfur dioxide is
    a. coal-fires electric generators
    b. automobile exhaust
    c. pesticide sprays
    d. mineral extraction

15. The ozone hole allowed an increase in
    a. atmospheric nitrogen concentration
    b. aerosol concentrations
    c. ultraviolet radiation
    d. chlorofluorocarbon concentrations

Reviewing the Facts

Fill in the blank with the word or words which most correctly completes each statement.

1. Carbon monoxide is a __________ pollutant.

2. Secondary pollutants are emitted __________ from identifiable sources.

3. Sulfuric acid is a _________ pollutant.

4. Photochemical reactions require strong _________.

5. Air pollution has become a problem in the past _____ years.

6. When air pollution episodes occur they _________ occur as the result of a large and rapid increase in the output of pollutants.

7. The largest single air pollutant in terms of the quantity emitted each year is _____________.

8. Surface temperature inversions most often develop in association with sinking air that characterizes slow moving centers of _________ pressure.

9. A solution is considered neutral when its pH is _______.

10. Temperatures within cities are generally _________ than in surrounding rural areas.

11. The wind speed in cities is generally __________ than in surrounding rural areas.

12. Although relative humidities are __________ in cities, cloud and fog frequencies are greater.

13. Although the atmosphere's carbon dioxide content has been rising, the outlook is that, in the very near future, the carbon dioxide content will _____________.

14. One possible result of the addition of carbon dioxide to the atmosphere is an _______ in cloud cover which, in turn, could partly offset a global temperature increase.

15. The addition to the atmosphere of trace gases such as ____________ and ___________ may cause a global temperature decline.

Developing and Applying Concepts

1. a. Does the complete combustion of coal and hydrocarbons threaten human health? Why do these fuels pose a pollution problem?
   b. State three products of incomplete combustion of hydrocarbons that are harmful to human health.
   c. State three products of the combustion of coal and hydrocarbons that are harmful to the environment. Specify how they are harmful.

2. State whether it is the concentration of the air's particles or volatiles that is increased by each of the situations described below:
   (a) a person sprays a can of air freshener
   (b) the gas tank of a car is filled at a gas station pump
   (c) a person combs their hair
   (d) a tractor plows a field
   (e) pesticides are sprayed onto crops by an airplane
   (f) sparks fly into the air from a campfire
   (g) a cloud is seeded
   (h) barbeque lighter fluid is poured on charcoal

3. Compare and contrast soot and fly ash.

4. Identify each as a primary or secondary air pollution:
   a. soot
   b. sulfur dioxide
   c. ozone
   d. acid rain
   e. nitrogen dioxide

5. Explain why ozone is harmful in the troposphere and helpful in the stratosphere.

6. You have measures the pH of four samples of rain water taken from four different locations. Arrange the samples in order of increasing acidity and point out which sample is the closest to "normal" rainwater.

7. Is pollution more likely to become concentrated on a windy or a calm bay? Explain.

8. List some of the types of air pollution that adversely affect vegetation and the condition of building materials. Specify the effect that each has.

9. List three sources of air pollution.

10. a. State four main areas which were addressed by the 1990 Clear Air Act amendment.
b. If this legislation exists, why are air pollution problems still occurring?

Thinking Critically about the Environment

1. Suppose you are a volunteer working with a group of people who live in a rainforest. A road is being planned to go through this locale. How would you advise them of the impacts of this road on their quality? Outline the arguments you will use to convince them.

2. Research the differences in the emissions produced by gasoline-burning and diesel fueled vehicles. Which, do you think, has the smallest impact on the atmosphere? Explain.

3. Explain the impact of the devices currently available to reduce indoor air pollution. Explain those you think are most important.

4. Radon testing has become a standard test for a home that is for sale. Would you insist on a radon test and how would the results influence your decision to buy the property?

Answers to Questions

Checking Vocabulary

1. a, 2. c, 3. c, 4. a, 5. b, 6. d, 7. b, 8. b, 9. a, 10. b, 11. d, 12. d, 13. a, 14. a, 15. c

Reviewing the Facts

1. primary, 2. indirectly, 3. secondary, 4. sunlight, 5. hundreds of, 6. rarely, 7. carbon monoxide, 8. low, 9. 7, 10. higher, 11. lower, 12. lower, 13. continue to increase, 14. increase, 15. methane, carbon dioxide

Developing and Applying Concepts

1. a. No. They only pose a problem when they are incompletely combusted.

b. benzene, carbon monoxide, and sulfur dioxide
c. water, carbon dioxide, and methane are some of the greenhouse gasses

2. (a) volatile
(b) volatile
(c) particle
(d) particle
(e) volatile
(f) particle
(g) particle
(h) volatile

3. Soot is mostly carbon, fly ash consists of non combustible materials that are not carbon. Both are particulates.

4. a. primary
b. primary
c. secondary
d. secondary
e. primary

5. Ozone absorbs ultraviolet radiation in the stratosphere and prevents it for reaching the surface where is would harm terrestrial plant and animal life. Ozone irradiates the respiratory membranes of animals when they breathe it and may be carcinogenic.

6. The order of increasing acidity is: 1,4,2,3; Sample 2 is closest to normal.

7. Pollution becomes more concentrated on a calm day as the movement of the sir tends to disburse and dilute any pollution, decreasing its concentration.

8. sulfur dioxide - plant destructor
ozone - impedes growth of leafy plants and may cause lung damage to animals
nitrogen dioxide - interferes with plant growth
acid rain - corrodes organic matter, rock, and metal

9. cooking vapors
aerosol sprays
house dust
hair
body odors
formaldehyde
radon

10. a. sulfur dioxide and acid rain
smog
volatile toxic compounds
ozone-destroying chemicals

b. Officials hesitate to close down businesses which may adversely affect the economy. Therefore when a business claims that it needs more time to meet pollution requirements, an extension is usually granted.

Chapter 20 Water Resources

Objectives

After you study this chapter, you should be able to:

- ☐ characterize water supply and demand
- ☐ differentiate between water withdrawal and consumption
- ☐ discuss the positive and negative aspects of water diversion
- ☐ describe problems developing from ground water withdrawal
- ☐ characterize methods for assessing hazards
- ☐ describe the types of pollutant sources

Chapter Outline

On a seperate sheet of paper, write a single sentence that summarizes the text for each part of the following outline.

- Water Supply and Demand
  - constant supply
  - types of demands
    - domestic
    - industrial
    - agricultural
  - water use
    - withdrawal
    - consumption
- Water Diversion
  - surface water diversion
    - types
    - effects
      - loss
      - salinization
      - silting
      - erosion
      - risk of disaster
      - recreation and aesthetic losses
      - ecological disruptions
  - ground water diversion
    - cone of depression
    - subsidence
    - salt water intrusion

- The Great American Desert
  - definition
  - examples
    - Colorado River
    - Los Angles Water Project
- Water and International Politics
  - principles
- Water Pollution
  - definition
  - sources
    - point
    - nonpoint
  - categories
- Sewage, Detergents, Fertilizers, and Pollution
  - types of impacts
- Toxic Pollutants: Assessment and Analysis
  - measurement of
    - threat
    - mobility
    - containment costs
  - cost-benefit analysis
- Ground Water Pollution
  - aquifer contamination
    - plume
  - treatment
    - elimination of source
    - monitoring
    - modeling
    - remediation
      - bioremediation
- Nuclear Waste Disposal
  - criteria
- Surface Water Pollution
  - case studies

Checking Vocabulary

Write a definition for each of the Key Terms at the end of this chapter. Check your knowledge of these terms by answering the following questions.

1. Problems with water supplies has occurred as the result of

   a. decrease in overall water supple

   b. increase in water demands

   c. changes in precipitation patterns

d. increase in sea water salinity

2. Domestic water use may be generally categorized as
   a. withdrawal
   b. consumption
   c. diversion
   d. salinization

3. Agricultural water use may be generally categorized as
   a. withdrawal
   b. consumption
   c. diversion
   d. salinization

4. The difference between water withdrawal and water consumption is
   a. the composition of the water returned
   b. the temperature of the water returned
   c. the location of the water returned
   d. the amount of water returned

5. The term describing the collection and transportation of surface and ground waters is
   a. withdrawal
   b. consumption
   c. diversion
   d. salinization

6. The drawbacks to surface water diversion do not include
   a. salinization
   b. loss by evaporation
   c. destruction of aquatic habitats
   d. hydroelectric generation

7. The major function of a dam is to
   a. generate electricity
   b. capture sediments
   c. provide recreational opportunities
   d. store water

8. The extraction of ground water
   a. produces a cone of depression
   b. increases contamination
   c. reduces total water supply
   d. prevents subsidence

9. Saltwater intrusion can be a problem
   a. when constructing a dam
   b. with coastal aquifers
   c. causing subsurface subsidence
   d. leading to excessive evaporation

10. More than half the population of North Americans get their drinking water from
   a. natural lakes and streams
   b. the melting of glacial ice
   c. coastal aquifers
   d. ground water sources

11. The original source of deep ground water was
   a. natural lakes and streams
   b. the melting of glacial ice
   c. infiltration of sea water
   d. coastal aquifers

12. Pollution originating from a specific site is termed
   a. natural pollution
   b. organic pollution
   c. point source pollution
   d. nonpoint source pollution

13. Fertilizers and pesticides are usually the cause of
   a. natural pollution
   b. organic pollution
   c. point source pollution
   d. nonpoint source pollution

14. Toxic inorganic compounds do not include
   a. pesticides
   b. mine wastes
   c. road salt
   d. industrial metals

15. The organic treatment of a contaminated aquifer is termed
   a. plume
   b. cost-benefit analysis
   c. point source pollution
   d. bioremediation

Reviewing the Facts

Fill in the blank with the word or words which most correctly completes each statement.

1. Water resources can be diminished by ____________ and ______________.

2. The ______________ of water is when it is used and returned to the earth locally.

3. Water consumption is a process which uses water and returns it _______ from its source.

4. In order of quantities, the three major categories of water usage are __________, __________, and __________.

5. Water diversion alters the ____________ of water.

6. Water may be lost from a reservoir by _________ and ___________.

7. Because streams deposit sediment in a reservoir, the result is ____________ downstream.

8. Ground water provides drinking water for ___________ of the population of North America.

9. A cone of depression is formed when ground water is extracted _________ than it can flow through an aquifer.

10. Ground water withdrawal can result in __________ the sinking of the earth's surface.

11. After the surface has subsided, the __________ is permanently reduced.

12. Two types of ground water occur in coastal areas, _________ and __________.

13. A _________ is a region that receives less than _______ centimeters of precipitation annually.

14. ________________ pollution arises from a specific site.

15. To eliminate a contaminant in a polluted aquifer, ____________ uses microorganisms.

Developing and Applying Concepts

1. Compare and contrast the two categories of water use.

2. Determine for each of the following water usages whether it is water withdrawal or consumption:

   a. industrial cooling
   b. toilet flushing

c. irrigation
d. car washing
e. drinking

3. List and explain some of the problems which may occur when water is diverted.

4. a. Why is ground water such a valuable natural resource?
b. Why is pumping ground water from a well like mining a natural resource?

5. The removal of ground water may create two problems. Explain these problems and steps that may be take to prevent or remedy each.

6. Define pollution and list eight categories of pollutants commonly found in surface and ground waters.

7. Compare and contrast point and nonpoint sources of water pollution.

8. Explain the steps taken in the process of aquifer remediation.

9. a. How are radioactive wastes different from other type of pollutants?
b. What steps must be taken to store these wastes safely?

10. Explain how cost-benefit analysis is used to assess the costs of water pollution control.

Thinking Critically about the Environment

1. What investigations would you think necessary before the construction of a dam as part of a water diversion project?

2. Find out where the drinking water in your community comes from. If from a reservoir, how is the water distributed and what type of backup system is there? If from a well, how many people are depending on the same aquifer?

3. Find out how landfills are constructed to prevent contamination of the ground water which may flow under them.

4. Make a list of the potential sources of point and nonpoint sources of water pollution in your community. What steps could be used to monitor and control each?

Answers to Questions

Checking Vocabulary

1. b, 2. a, 3. b, 4. c, 5. c, 6. d, 7. d, 8. a, 9. b, 10. d, 11. b, 12. c, 13. d, 14. a, 15. d

Reviewing the Facts

1. depletion, pollution, 2. withdrawal, 3. far, 4. domestic, industrial, agricultural, 5. distribution, 6. evaporation, leaking, 7. erosion, 8. one half, 9. faster, 10. subsidence, 11. porosity, 12. fresh, salty, 13. desert, 14. point source, 15. bioremediation

Developing and Applying Concepts

1. Withdrawal is any process that uses water and returns it close to the point of its origin. Water consumption is a process which uses water and returns it far from its source.

2.
   a. withdrawal
   b. withdrawal
   c. consumption
   d. withdrawal
   e. withdrawal

3. Some of the problems which occur with water diversion projects are:
   a. water loss from evaporation and leakage
   b. salinization resulting from concentration of natural dissolved minerals
   c. silting from stream deposits accumulating in reservoirs
   d. erosion of areas downstream of a reservoir
   e. increased risk of disaster from dam breaks
   f. aesthetic losses from the flooding of areas behind dams
   g. ecological impacts from the disruption of natural river systems.

4.
   a. Ground water is valuable because it is abundant, it is stored underground and may b available during times of drought, and it is available in many, otherwise arid environments.
   b. Since ground water moves slowly, it takes a long time to become replenished, and the rate of extraction may exceed the rate of replenishment.

5. The removal of ground water may cause subsidence as the aquifer particles move to fill emptied pore spaces. This problem is irreversible. In coastal areas ground water removal can lead to saltwater intrusion. This may be prevented by returning water to the ground after it has been used.

6. Pollution is the reduction of the quality of a resource through the introduction of impurities. Eight common water pollutants are; sewage, disease organisms, plant nutrients, industrial organic compounds, toxic inorganic compounds, sediment, radioactive materials, and heat.

7. Point source pollution arises from a specific site and is much easier to identify and control. Nonpoint source pollution is generated over a broad area. This results in very low concentrations

in any particular source location, but as water moves through, the concentrations may rise to dangerous levels. Nonpoint sources are much more difficult to control.

8. The steps in the process of aquifer remediation are;

a. elimination of the pollution source
b. monitoring to assess the movement rates and directions of the contaminant plume
c. develop a predictive model for the aquifer
d. cleaning of the aquifer using a method appropriate for the aquifer and the pollutant

9. a. Radioactive wastes cannot be decontaminated by any known process. They will only decay naturally over a time specific for each isotope.
b. The wastes must be stored in a location safe from disruption by volcanoes, earthquakes, landslides, floods, and ground water so that they are not spread out into the environment before they decay.

10. A cost-benefit analysis compares the costs of pollution control measures with the costs of externalities, the costs encountered with the spread of pollution. The result is a scale with a pristine environment (absolutely zero pollution) being very expensive, in dollars, and an uncontrolled environment (high pollution levels) being much cheaper. The externalities, non-dollar costs, make the analysis a basis for decision makers.

Chapter 21 Geologic Resources

Objectives

After you study this chapter, you should be able to:

- ☐ characterize the two major categories of geologic resources
- ☐ list and describe the five types of processes which create mineral resources
- ☐ discuss the origin and forms of fossil fuels
- ☐ list the positive and negative aspects on nuclear power
- ☐ describe alternate energy resources

Chapter Outline

On a seperate sheet of paper, write a single sentence that summarizes the text for each part of the following outline.

- Geologic Resources
    - types
        - mineral resources
            - ore
            - reserves
            - metals
            - nonmetallic
            - nonrenewable
        - energy resources
            - fossil fuels
            - nuclear fuels
            - alternates
            - nonrenewable & non-recyclable
- Ore Formation
    - magmatic processes
    - hydrothermal processes
        - vein deposits
        - disseminated deposits
            - sea floor deposits
    - sedimentary processes
        - placer deposits
        - precipitation
        - weathering deposits
- Mineral Reserves
    - sizes
        - discovery

ore grades
Coal
formation
Mines and Mining
types
underground
surface
Petroleum and Natural Gas
formation
source rock
oil trap
reservoir
extraction
transportation
refining
natural gas
tar sands
oil shale
Nuclear Fuels and Reactors
nuclear fission
branching chain reaction
Energy Strategies
sources
use
Alternative Energy Resources
solar energy
wind energy
geothermal energy
hydroelectric energy
biomass energy
sea energy
hydrogen fuel
conservation
strategies

Checking Vocabulary

Write a definition for each of the Key Terms at the end of this chapter. Check your knowledge of these terms by answering the following questions.

1. Useful rocks and minerals are called
   a. mineral resources
   b. energy resources

c. renewable resources
d. alternative resources

2. A rock enriched in a useful mineral is termed a(n)
a. metal
b. ore
c. fuel
d. fossil

3. Magmatic processes produce ores by
a. hydrothermal separation
b. mineral evaporation
c. crystal settling
d. nuclear fission

4. Disseminated ore deposits are produced by
a. magmatic separation
b. precipitation
c. hydrothermal processes
d. weathering processes

5. Manganese nodules are produced by
a. magmatic separation
b. precipitation
c. hydrothermal processes
d. weathering processes

6. Sedimentary ore deposits are produced by
a. sorting or evaporation
b. very hot ground water
c. crystal settling
d. nuclear fission

7. Bauxite, the source of aluminum, forms as the result of
a. magmatic separation
b. precipitation
c. hydrothermal processes
d. weathering processes

8. Energy resources differ from mineral resources as they are
a. nonrenewable
b. more concentrated
c. nonrecyclable

d. more rare

9. The global distribution of mineral resources is
    a. random
    b. uniform
    c. uneven
    d. unknown

10. Mineral reserves may increase by
    a. mining
    b. discovery
    c. refining
    d. extraction

11. Fossil fuels form from
    a. magmatic processes
    b. very hot groundwater
    c. sedimentary separation
    d. plant and animal remains

12. One of the problems with the use of coal is
    a. emission of air pollutants
    b. limited supplies
    c. difficulty of extraction
    d. lack of application

13. The material in which oil originally forms is called the
    a. oil trap
    b. petroleum reservoir
    c. salt dome
    d. source rock

14. The major drawback to nuclear fission energy is
    a. the production of heat
    b. the lack of fuel supplies
    c. the creation of chain reactions
    d. the disposal of waste

15. Alternative energy resources may be characterized as
    a. non-economic
    b. renewable
    c. non-polluting
    d. recyclable

Reviewing the Facts

Select one of the two italicized words in parentheses to correctly complete each of the following statements.

1. Water consumption is a problem in the development of *(petroleum, oil shale)*.

2. A reservoir rock is *(permeable, impermeable)*.

3. Bauxite is an example of a(n) *(residual, evaporite)* deposit.

4. Fossil fuels are *(energy, mineral)* resources.

5. The processing of oil shale *(is, is not)* presently economical.

6. Banded iron formations formed when the *(oxygen, water vapor)* content of the earth's atmosphere increased.

7. The ore deposits of the Bushveld were formed by *(hydrothermal solutions, crystal settling)*.

8. An example of a nonmetallic resource is *(platinum, sand)*.

9. At temperatures of 100° to 150° C, *(petroleum, natural gas)* is formed from organic matter.

10. Gold can be found in *(evaporite, placer)* deposits.

11. Ceramic objects and bricks *(are, are not)* both made of clay.

12. Nuclear power plants *(have, have not)* been built since 1981.

13. A source rock is *(permeable, impermeable)*.

14. Coal is formed from *(oil shale, peat)*.

15. A cap rock is *(permeable, impermeable)*.

Developing and Applying Concepts

1. a. Can natural gas and petroleum form from the same source rock? Explain.
   b. Which of the following situations would be more likely to form a petroleum reservoir:
      (1) a rise in sea level in which a near shore lagoon is overlain by beach sand, or
      (2) a drop sea level in which a near shore lagoon is exposed on dry land?
   c. Why is it useful for geologists to understand paleoenvironments when they are exploring for petroleum reservoirs?

2. a. Are fossil fuels nonrenewable resources? Explain.
b. List the following fuels in order of the length of time supplies are predicted to last at the current rates of consumption:
(1) coal, (2) petroleum, (3) natural gas
c. Why is coal mining an environmental problem?
d. Why is it that, despite the Surface Mining Control and Reclamation Act, there are thousands of mine sites that have not been restored?

3. Summarize the problems associated with petroleum extraction, transport, and refining.

4. Oil is not found in Precambrian rocks. Explain why this is so.

5. a. How important is nuclear fission as an energy source in the United States?
b. Why isn't nuclear fission currently growing as a source of energy?
c. Why isn't nuclear fusion a source of energy in the Unites States?

6. a. Is it possible to predict where an ore of a certain type is likely to be found? Explain.
b. A geologist found an ore deposit containing 0.22% by weight of uranium, 0.20% by weight of lead, and 1.50% by weight of zinc. Use Table 21-2 to determine if it would be economically feasible for this ore to be mined. Explain.

7. Why would spreading centers and subduction zones be likely places to find metallic ore deposits?

8. Explain two sedimentary processes which may form ore deposits.

9. Specify the type of process that could form ore deposits in each of the following:
(1) joints of igneous rocks
(2) ancient soils
(3) batholiths
(4) point bar deposits

10. Explain the process which forms bauxite.

Thinking Critically about the Environment

1. Determine how much oil is obtained from rocks of each geologic period and make a bar graph to demonstrate the results. Explain the trends you see.

2. Research the various types of anti-pollution devices that are used in any one area of energy production. Contact a company involved in this area, such as petroleum refining or mining, and find out what proportion of income is spent on such devices.

3. Make a poster display of common household items and the metallic or nonmetallic mineral

resources that they contain.

4. Construct a flow diagram to illustrate energy input and outflow of processing a fossil fuel. Include measures taken to prevent pollution.

5. Find out where the electrical energy used by your community comes from and how it is produced.

6. Make a list of the strategic minerals of the United States, the minerals which are important to our society but must be imported. What countries serve as major suppliers of these resources?

7. Research the history and environmental impact of hydraulic mining in California.

Answers to Questions

Checking Vocabulary

1. a, 2. b, 3. c, 4. c, 5. c, 6. a, 7. d, 8. c, 9. c, 10. b, 11. d, 12. a, 13. b, 14. d, 15. b

Reviewing the Facts

1. oil shale, 2. permeable, 3. residual, 4. energy, 5. is not, 6. oxygen, 7. crystal settling, 8. sand, 9. natural gas, 10. placer, 11. are, 12. have not, 13. impermeable, 14. peat, 15.impermeable

Developing and Applying Concepts

1. a. Yes. Natural gas and petroleum can form from the same source rock because they both form from the decay of organic matter. Petroleum forms when organic matter is heated from 50° to 100° C. Natural gas forms when crude oils is heated from 100° to 150° C.
b. (1) The lagoon sediments will serve as the source rock and the oil will be able to migrate into the beach sands.
c. It is useful for geologists to understand paleoenvironments when they are looking for petroleum reservoirs because it helps them to predict where rocks of certain types may be found.

2. a. Yes, fossil fuels are nonrenewable resources because there is only a fixed amount on earth. Although new deposits are forming today, the process is too slow to replenish supplies.
b. (2) petroleum, (3) natural gas, (1) coal
c. With underground coal mines collapse and mine spoil disposal are problems. Surface coal mines leave huge holes and piles of mine spoil. In both cases, when coal is exposed to air, sulfuric acid may form resulting in acid mine drainage. When coal is burned, the sulfur reacts to form hydrogen sulfide and sulfuric acid which leads to the production of acid precipitation.

d. The law is not being adequately enforced.

3. The main problems associated with petroleum extraction, transportation, and refining are accidents. During petroleum extraction, broken pipes, excess pressures, and difficulties in capping new wells lead to blowouts, spills, and fires. Tanker accidents during transport pollute aquatic environments. Air pollution may result from refining processes.

4. Many Precambrian rocks have been metamorphosed an any oil which was in them has been eliminated.

5. a. Nuclear fission provides 7.6% of the energy used in the United States.
b. The theat of accidents and the problems associated with the disposal of radioactive wastes have caused the construction of nuclear reactors to come to a halt.
c. The technology for the economical production of energy by nuclear fusion does not yet exist.

6. a. Yes, certain metals may be associated with certain types of igneous rocks.
b. Yes, this ore could be mined for uranium and lead. The zinc could not be extracted economically on its own but might be extracted along with the other two.

7. Spreading centers and subduction are areas of high igneous activity. Hydrothermal deposits will be abundant in both because fluids enriched in metallic minerals will rise from underground magma bodies. Subduction zones are also characterized by great volumes of plutonic rocks.

8. Heavy minerals are the first particles to be deposited by water when its velocity is decreased. Since this occurs on the inside of meanders, placer deposits form there. Precipitation is the other sedimentary process that may cause ore formation. Precipitation of minerals can occur by evaporation of the water or by other chemical changes, such as oxidation.

9. (1) hydrothermal
(2) weathering (residual deposits)
(3) magmatic (crystal settling)
(4) sedimentary (placer deposits)

10. Bauxite is formed by weathering processes when rainwater leaches ions from the upper layers of a soil, carrying them to lower layers.

Unit VII: Astronomy

Chapter 22 Motion in the Heavens

Objectives

After you study this chapter, you should be able to:

•☐ describe the differences between the geocentric and heliocentric universe

•☐ discuss the problem that the heliocentric theory solved

•☐ list the differences between lunar and solar eclipses

•☐ describe the techniques used to study space

•☐ characterize emission and absorption spectra

•☐ discuss the impacts of the Doppler effect

Chapter Outline

On a seperate sheet of paper, write a single sentence that summarizes the text for each part of the following outline.

- Motions of the Heavenly Bodies
  - cyclicly
    - sun
    - moon
  - constellations
  - planets
    - retrograde motion
- Aristotle and the Earth-Centered Universe
  - geocentric
  - celestial sphere
  - parallax
- Renaissance and the heliocentric Solar System
  - ideas and contributions of:
    - Copernicus
    - Brahe
    - Kepler
    - Galileo
    - Newton
- Motions of the Earth and the Moon
  - rotation
  - revolution

precession
moon's motions
phases
waxing
waining
eclipses
solar
umbra
penumbra
lunar
Modern Astronomy
telescopes
optical
refracting
reflecting
other wavelengths
spectra
emission
absorption
Doppler measurements
color shifts
spacecraft

Checking Vocabulary

Write a definition for each of the Key Terms at the end of this chapter. Check your knowledge of these terms by answering the following questions.

1. The turning of the earth on its axis is called
   a. revolution
   b. inertia
   c. parallax
   d. rotation

2. Retrograde motion refers to
   a. the backward motion of the planets
   b. general star movement
   c. the spiraling motion of the earth in its orbit
   d. the wobbling of the earth's axis

3. When the moon is positioned between the earth and the sun blocking out sunlight, it is called a
   a. lunar eclipse
   b. terrestrial eclipse
   c. solar eclipse

d. coronal eclipse

4. When the moon appears fully light by the sun as a result of its position whit respect to the earth it is called a
   a. full moon
   b. crescent moon
   c. quarter moon
   d. new moon

5. Planets
   a. have a fixed position with respect to the stars
   b. always appear to move in the same direction with respect to the stars
   c. usually move in a given direction with respect to the stars
   d. move erratically with respect to the stars

6. According to the heliocentric model of the universe, the
   a. sun is the center of the solar system
   b. earth is the center of the universe
   c. earth is the center of the solar system, but not of the universe
   d. solar system has no center

7. A celestial sphere is
   a. another term for the earth's atmosphere
   b. a planet
   c. a star
   d. an imaginary crystal sphere which supports celestial objects

8. When the moon appears as a tiny curved sliver it is called a
   a. full moon
   b. crescent moon
   c. quarter moon
   d. new moon

9. When the moon is positioned behind the earth with respect to the sun with the earth blocking out the sun's light it is termed a
   a. lunar eclipse
   b. terrestrial eclipse
   c. solar eclipse
   d. coronal eclipse

10. When the moon appears dark because it is between the earth and the sun it is called a
   a. full moon
   b. crescent moon

c. quarter moon
d. new moon

11. The penumbra is
a. the dark part of the shadow formed during a solar eclipse
b. the light part of the shadow formed during a solar eclipse
c. the sun's outer atmosphere visible only during an eclipse
d. a reflective halo which appears around the moon during an eclipse

12. The ability of the eye to distinguish details of an object is called
a. resolution
b. parallax
c. revolution
d. absorption

13. According to the geocentric model of the universe, the
a. sun is at the center of the universe
b. universe has no center
c. sun is the center of the soar system, but not the universe
d. earth is the center of the universe

14. The apparent change of position of an object due to the change in position of the observer is called
a. retrograde motion
b. parallax
c. gravity
d. inertia

15. The corona is
a. the dark part of the shadow formed during a solar eclipse
b. the light part of the shadow formed during a solar eclipse
c. the sun's outer atmosphere visible only during an eclipse
d. a reflective halo which appears around the moon during an eclipse

Reviewing the Facts

Determine whether the following statements are true or false. If the statement is false, change the underlined word so that it is correct.

1. A telescope that uses a mirror to gather light is called a reflecting telescope.

2. A solar eclipse is visible if the observer stands in the penumbra of the moon's shadow.

3. Fixed groups of stars named by the ancients are called constellations.

4. The first person to show that the universe was "blemished" was Newton.

5. The human eye can detect about 99% of the electromagnetic spectrum.

6. Another term for a dark-line spectrum is an emission spectrum.

7. The moon rotates once on its axis each time it revolves around the earth.

8. The ancients did observe stellar parallax.

9. The tendency of a object to resist a change in motion is called inertia.

10. Astronomers may study the universe using all of the wavelengths of the electromagnetic spectrum.

11. The same side of the moon is not always visible from earth.

12. Brahe was the scientist who first proposed that the planets follow elliptical orbits.

13. One complete cycle of lunar phases occurs every 27.5 days.

14. Lunar eclipses are more common than solar eclipses.

15. Newton proposed that gravity keeps celestial bodies in their orbits.

Developing and Applying Concepts

1. Why were planets called "wanderers" by the ancients?

2. What contributions did Brahe and Kepler make to science?

3. a. Contrast the geocentric and heliocentric models of the Universe and the Solar System.
b. Why did it take so long for the geocentric theory to be replaced by the heliocentric theory?

4. a. Galileo observed that Venus has phases like the moon. Using a diagram of the solar system to determine whether or not any other planets have phases. Explain.
b. Imagine that you are looking at a planet through a telescope and you notice that every 10 earth days a certain feature reappears on the surface of the planet. What can you infer from this about the planet?

5. What observation of the night sky does the precession of the earth's axis affect?

6. a. If a full moon rises at sunset (6 pm) and sets at dawn (6 am), at what time of the night will it be highest in the sky?
b. Order the following phases of the moon so that they make a complete cycle. Begin with a new moon: new, crescent (right side lighted), third quarter, first quarter, old crescent (left side lighted), new gibbous (right 3/4 lighted), old gibbous (left 3/4 lighted).

7. Why are lunar eclipse more common than solar eclipses?

8. Describe three limitations of viewing celestial objects with the naked eye.

9. a. Describe how optical and other types of telescopes enhance the observation of celestial objects.
b. Why is the size of a reflecting telescope limited?
c. How can the Keck telescope be four times larger than the one at the Palomar Observatory?
d. What advantage does the Hubble Telescope have over both the Palomar and Keck Telescopes?

10. Compare and contrast emission and absorption spectra.

Thinking Critically about the Environment

1. It has been suggested that the orbital energy of the moon can be harnessed by construction a generator that makes use of tidal energy. Research this possibility. Has a tidal power generator been constructed? Could all coastal areas be serviced by such a generator? Could inland areas be serviced? What problems would be encountered?

2. Celestial objects are not the only things that can be viewed using the full range of wavelengths in the electromagnetic spectrum. Satellites have viewed earth in wavelengths that fall outside the range of visible light and have obtained some extremely interesting and useful information. Prepare a report about the types of radiation that yield important information about the earth. What earth characteristics are made "visible" in this manner?

Answers to Questions

Checking Vocabulary

1. d, 2. a, 3. c, 4. a, 5. c, 6. a, 7. d, 8. b, 9. a, 10. d, 11. b, 12. a, 13. d, 14. b, 15. c

Reviewing the Facts

1. true, 2. umbra, 3. true, 4. Galileo, 5. one millionth of one percent, 6. absorption, 7. true, 8. did not, 9. true, 10. true, 11. is, 12. Kepler, 13. 29.5, 14. true, 15. true

Developing and Applying Concepts

1. The planets were called "wanderers" by the ancients because they moved fairly rapidly across the sky relative to the fixed background of stars.

2. Brahe carefully recorded the movements of the planets. Kepler analyzed those movements and concluded that the planets revolve around the sun in elliptical orbits.

3. a. In the geocentric model, the earth is the center of the universe. The moon, sun, planets, and stars are carried around the earth on crystalline spheres. In the heliocentric model, the sun is the center of the solar system. The earth is a planet, which like the others, orbit the sun.
b. The geocentric theory was supported by philosophical and religious views during the time the theory was accepted. The testing of ideas with experimentation and observations was not done until Galileo's time. He convinced people that testing and observations were more reliable than faith.

4. a. Both the interior planets, Venus and Mercury, have phases because they are between the earth and the sun. The outer planets do not have phases. Observers on earth always see the outer planets fully lighted since the earth is always located between then and the sun.
b. If every 10 days you notices the same feature on a planet's surface you could conclude that either the planet has a moon which passes in front of it or it rotates once every 10 days.

5. The precession or wobbling of the earth's axis affects the position of the Pole Star. Earth's axis currently points toward the North Star. Gradually this will change and the axis will point toward a different star.

6. a. 12 pm (midnight)
b. new, new crescent, first quarter, new gibbous, full, old gibbous, third quarter, old crescent

7. Lunar eclipses are more common than solar eclipses because the earth is larger than the moon and its shadow is more likely to cover the entire lunar surface.

8. 1) The naked eye only detects visible light which is one millionth of one percent of the electromagnetic spectrum
2) The naked eye doesn't collect much light
3) The naked eye has poor resolution.

9. a. Telescopes can collect light for long periods of time if photographic film is used. Also, the size of the light collecting surface can be much larger than the human eye's retina. Lastly, lenses and mirrors can enhance resolution.

b. The size of a reflecting telescope is limited because a mirror sags under its own weight if it much larger than 600 centimeters.
c. The Keck telescope is four times larger then the Palomar telescope because it is composed of 36 smaller mirrors that are aligned and focused with a computer.
d. The Hubble telescope is orbiting earth above its atmosphere and does not have to contend with distortions caused by the atmosphere.

10. Absorption spectra appear as dark lines or missing wavelengths in the continuous spectrum of the electromagnetic radiation. Emission spectra consist of individual bright lines which represent single wavelengths of the electromagnetic spectrum.

Chapter 23 Planets and their Moons

Objectives

After you study this chapter, you should be able to:

- ☐ list the planets with respect to their distance from the sun
- ☐ characterize the major differences between the terrestrial and Jovian planets
- ☐ discuss the relationship of asteroids, meteorites, and meteoroids
- ☐ describe how a comet changes as it moves through the solar system

Chapter Outline

On a seperate sheet of paper, write a single sentence that summarizes the text for each part of the following outline.

Overview of the Solar System
  formation
  solar wind
  terrestrial planets
  Jovian planets
Mercury
  size
  atmosphere
  surface features and geology
Venus
  size
  atmosphere
  surface features and geology
The Moon
  formation
  size
  atmosphere
  surface features and geology
Mars
  size
  atmosphere
  surface features and geology
  moons
  life?
Jupiter
  size
  atmosphere

surface features and geology
moons

Saturn
- size
- atmosphere
- surface features and geology
- rings
- moons

Uranus and Neptune
- size
- atmosphere
- surface features and geology
- rings and moons

Pluto
- size
- atmosphere
- surface features and geology
- moon

Asteroids, Comets and Meteoroids
- asteroids
  - formation
  - composition
  - location
- comets
  - structure
  - orbits
- meteoroids
  - types
  - compositions

Checking Vocabulary

Write a definition for each of the Key Terms at the end of this chapter. Check your knowledge of these terms by answering the following questions.

1. A celestial object which is composed of ice mixed with bits of silicate rock, metals, and frozen crystals, is a(n)
   a. comet
   b. asteroid
   c. meteoroid
   d. moon

2. The Jovian planet are
   a. Mercury, Venus, Earth, and Mars

b. Mars, Jupiter, Saturn, and Neptune
c. Jupiter, Saturn, Uranus, and Neptune
d. Uranus, Neptune, Pluto, and Saturn

3. The least dense planet in the solar system is
a. Saturn
b. Jupiter
c. Neptune
d. Uranus

4. The planet most distant from the sun is
a. Pluto
b. Neptune
c. Mercury
d. Jupiter

5. The Galilean moon which has active volcanoes is
a. Callisto
b. Europa
c. Ganymede
d. Io

6. The terrestrial planets are
a. Mercury, Venus, and Earth
b. Jupiter, Saturn, Uranus, and Neptune
c. Mars, Jupiter, Saturn, Uranus, and Neptune
d. Mercury, Venus, Earth, and Mars

7. Mars does not
a. experience seasonal dust storms
b. have polar ice caps
c. have active volcanoes
d. show signs of water erosion

8. Methane exists in three states on the natural satellite
a. Europa
b. Titan
c. Charon
d. Io

9. Small, round grains composed of olivine and pyroxene that are formed in meteorites are called
a. maria
b. chondrules

c. asteroids
d. nuclei

10. The planet which is about the size of the earth's moon, has a heavily cratered surface, no atmosphere, and is tectonically active is
a. Mars
b. Venus
c. Pluto
d. Mercury

11. The largest planet of the solar system is
a. Saturn
b. Jupiter
c. Neptune
d. Uranus

12. The Great Red Spot is
a. an ancient volcano on Mars
b. a cyclonic storm on Jupiter
c. a moon which orbits Saturn
d. turbulence in the atmosphere of Neptune

13. A shooting star si a(n)
a. meteoroid
b. asteroid
c. meteor
d. meteorite

14. Liquid metallic hydrogen
a. is formed by great pressure which frees electrons from atoms
b. forms the core of Mars
c. is created by the solar wind
d. is found in the atmosphere of Neptune

15. The bright outer sheath of a comet is called the
a. tail
b. maria
c. coma
d. nucleus

Reviewing the Facts

Select the letter of the choice which most correctly completes each of the following statements.

1.The earth's moon and Mercury are similar in that both
a. rotate once in their axis each time they revolve
b. have heavily cratered surfaces
c. were mapped by Galileo
d. have highly reflective regions at their poles

2. The planets formed by a process called
a. gravitational coalescence
b. blob tectonics
c. meteorite bombardment
d. lithospheric accumulation

3. Moon rocks were collected and brought back to earth by astronauts of the
a. Apollo program
b. Magellan program
c. Voyager program
d. Viking program

4. Uranus and Neptune differ from Jupiter and Saturn in that they have
a. liquid surfaces and interiors
b. dense atmospheres
c. interiors of rock and metals
d. tilted magnetic fields

5. The main reason why the terrestrial planets are different from the Jovian planets is that the
a. the original composition of the cloud of dust and gas from which the solar system formed varied from the center to the outer regions
b. the sun's heat boiled off the lighter gasses from the inner planets of the solar system in its early stages of formation
c. gravitational attraction of the Jovian planets was large enough to draw the lighter elements to the outre part of the solar system
d. lighter elements migrated into the solar system from space and were retained in the atmospheres of the first planets they encountered

6. Venus was mapped by a space probe called
a. Voyager
b. Mariner
c. Viking
d. Magellan

7. Mercury
a. has extremes of temperatures which range from 450° to -175°
b. has temperatures all over the planet high enough to melt lead

c. is extremely cold all over the planet
d. has moderate temperatures similar to earth

8. The only planetary moon known to have an atmosphere is
a. Titan
b. Callisto
c. Ganymede
d. Io

9. The moon
a. never melted during its history
b. is entirely rigid and solid
c. is of the same rock type throughout
d. is tectonically inactive

10. Volcanism on one of Jupiter's moons is thought to be the result of
a. plate tectonics
b. radioactive heating
c. tidal forces
d. heat generated by meteorite bombardment

11. The asteroid belt is located between the orbits of
a. Mercury and Venus
b. Earth and Mars
c. Mars and Jupiter
d. Jupiter and Saturn

12. The clouds of Venus rain
a. methane
b. liquid carbon dioxide
c. sulfuric acid
d. water

13. The geology of Mars is most similar to that of
a. Mercury
b. Venus
c. Earth
d. Earth's moon

14. Pluto is
a. a typical Jovian planet
b. a typical terrestrial planet
c. like Earth's moon

d. like Neptune's moons

15. Saturn's rings are composed of
   a. gas
   b. dust
   c. gas and dust
   d. gas, dust, and ice

Developing and Applying Concepts

1. a. Summarize the basic differences between the terrestrial and Jovian planets.
   b. What caused these differences?
   c. Why is Pluto considered an "oddball"?

2. a. Compare Mercury and the Moon.
   b. Why is Mercury the densest of all planets?
   c. Why is the surface of Mercury so old?

3. a. Interpret the internal processes of Venus from its surface characteristics.
   b. Venus lacks a magnetic field. What does this suggest about its interior structure?

4. a. Relate the types of rock found on the moon to its topography.
   b. List two periods of time in which heavy swarms of meteorites bombarded the moon.
   c. Maria are far more common on the side of the moon which faces the earth than on the side which faces away. State a hypothesis based on what you have learned about Jupiter's moon Io, that explains this phenomenon.
   d. Compare and contrast the characteristics and mode of formation of the regoliths on the earth and the moon.
   e. Why do meteorites have a grater impact on the moon than they do on the earth?
   f. Why do meteor craters last longer on the moon's surface than they do on earth?
   g. When a meteor crater is formed, a layer of ejected material settles around the crater. Explain how this feature can be used to determine the relative ages of the craters.
   h. How would pictures of the surfaces of the earth and the moon taken 400 million years ago compare to how they look today?
   I. Given the rate at which the earth's surface changes, what is wrong with the idea that the moon was torn from the earth, creating the Pacific Ocean basin as a result?

5. a. The Martian day is about the same length as an earth day and both planets are tilted about the same amount on their axes. Why, then, does Mars have a greater extreme of temperatures on its surface than the earth?
   b. Why are the surfaces of Mercury and the moon are much older than the earth's surface?
   c. Why has Mars been targeted as being the most likely other planet for the development of life?

6. Why does Io exhibit active volcanism and Jupiter's other moons do not?

7. Describe Saturn's rings and explain their formation.

8. Compare and contrast Uranus and Neptune.

9. a. Why is it thought that Pluto and Charon might be escaped moons of Neptune?
b. What do the earth and Moon have in common with Pluto and Charon?

10. Compare and contrast asteroids and meteoroids.
b. What is the origin of most meteoroids?
Some meteoroids found in Antarctica are thought to have originated from the Moon add Mars. Infer how this could have occurred.

Thinking Critically about the Environment

If you could travel to any other planet or planet's moon to establish a human settlement what would be the criteria for your choice of location? Why?

Answers to Questions

Checking Vocabulary

1. a, 2. c, 3. a, 4. a, 5. d, 6. d, 7. c, 8. b, 9. b, 10. d, 11. b, 12. b, 13. c, 14. a, 15. c

Reviewing the Facts

1. b, 2. a, 3. a, 4. d, 5. b, 6. d, 7. a, 8. a, 9. d, 10. c, 11. c, 12. c, 13. b, 14. d, 15. d

Developing and Applying Concepts

1. a. The terrestrial planets are small and rocky with metallic cores. The Jovian planets are much larger and have thick atmospheres, liquid surfaces and interiors and a relatively small, solid core.
b. The differences between the terrestrial and Jovian planets resulted when the sun formed. Its heat boiled off the lighter elements from the inner planets, leaving mostly rocky substances behind. The more distant Jovian planets accumulated large amounts of these lighter elements.
c. Pluto is very small when compared to the Jovian planets. Its composition and structure are thought to be more similar to a moon of Neptune.

2. a. Mercury and the Moon both have heavily cratered and old surfaces. They are about the same size and are tectonically dead. They also both lack atmospheres and have extremes of temperatures. The rates of rotation are slow - Mercury rotates three times for every two revolutions and the Moon rotates once per revolution.

b. Since Mercury was closest to the sun during the formation of the solar system, a higher percentage of volatiles were boiled away an only the heaviest elements were left behind.
c. The surface of Mercury is old because it has been tectonically inactive for more than 4 billion years.

3. a. Volcanic mountains have been identified on Venus. The surface of Venus is not heavily cratered, indicating that it is covered with relatively recent lava flows. It has several mountain chains, large crustal fractures and deep canyons. However, it lacks steep scarps, transform faults, or features analogous to oceanic trenches. It has been suggested that upwellings of hot rock from the planet's interior are the source of volcanism and have created non-volcanic mountains by uplifting. The fractures and deep canyons could be the result of stretching forces produced by the uplift.
b. The lack of a magnetic field in Venus suggests that it does not have a molten interior similar to the earth's outer core.

4. a. Basalt is the type of rock that the flat maria areas are made of. Anorthosite is the rock type of the highlands. The maria is not heavily cratered, the highlands are.
b. The first bombardment occurred more than 4.4 billion years ago, the second was between 4.2 and 3.9 billion years ago.
c. The earth's tidal forces on the moon probably caused greater melting on the side of the moon which faces the earth.
d. The regoliths of both the earth and the moon contain pieces of broken rock. The earths regolith contains organic matter and is formed by both chemical and mechanical weathering. The moon's regolith is formed only by mechanical processes as particles from space impact on the lunar surface.
e. Meteorites have a grater impact on the moon than they do on the earth because to moon has no atmosphere to slow them and decrease their size.
f. Meteorite craters last longer in the moon because the only erosion which occurs on the moon is the result of incoming particles from space and the moon's gravity.
g. A crater which is covered by ejecta from another crater is the older.
h. The moon's picture would look almost exactly as it is today. The earth's picture would be unrecognizable.
i. Any "scar" left by the separation of the moon from the earth more than 4 billion years age would have been obliterated by the actions of plate tectonics.

5. a. Mars does not have much atmosphere to facilitate heat transfer and equalize surface temperatures.
b. Mercury and the moon are tectonically inactive and have been for billions of years. Also, neither has an atmosphere to wear down surface rocks.
c. Early observations of Mars revealed patterns which changed with time. The existence of polar ice caps indicated that water might be present on the surface. Although these indications have not held up with the increasing knowledge of Mars, it is still thought that frozen water exists on Mars and that life may have once developed there.

6. Io is closest to Jupiter and therefore experiences the strongest tidal forces.

7. Saturn's rings are 10 to 25 meters thick and 425,000 kilometers wide. They are composed of dust, rock, and ice particles which range from a few meters to a few ten-thousandths of a centimeter in diameter. The rings may be fragments of a moon that never formed or the remains of one which broke apart after its formation.

8. Uranus and Neptune have low densities. They have dense atmospheres, liquid surfaces and interiors, and solid rock and metal cores. Both planets have a higher proportion of rock than Jupiter and Saturn. Their atmospheres are mainly hydrogen and helium with smaller amounts of carbon, nitrogen, and oxygen. The liquid layers are thought to be methane, ammonia, and possibly water. Both planets have a tilted axis.
The two planets differ in that the atmosphere of Uranus is calm and that of Neptune is turbulent. Uranus has 15 moons, Neptune has 8 moons.

9. a. Pluto and Charon are unlike the Jovian planets in size and composition. Pluto is similar in size and density to Neptune's moon, Triton, and is thought to be composed of frozen methane.
b. Both the Moon and Charon are about the same size as the planet they orbit, when compared to the size difference between other moons and their planets.

10. a. Asteroids and meteoroids have similar compositions. Asteroids are generally larger then meteoroids and most orbit the sun between Jupiter and Mars.
b. Meteoroids are formed when asteroids collide with one another and break apart.
c. Pieces of the Moon and Mars could have been dislodged by a meteorite impact. If energy was sufficient that could have been thrown out of the gravitational field of these bodies and hurled into space. Later they were captured by the earth's gravitational field and pulled down to the surface.

Chapter 24 Stars, Space, and Galaxies

Objectives

After you study this chapter, you should be able to:

- ☐ describe how distances beyond the solar system are measured
- ☐ explain the formation of stars
- ☐ discuss the structure of the sun
- ☐ describe what the Hertzsprung-Russell diagram illustrates
- ☐ list the types of galaxies
- ☐ discuss the origin of the universe

Chapter Outline

On a seperate sheet of paper, write a single sentence that summarizes the text for each part of the following outline.

- Measuring Distances in Space
  - light year
  - parsec
- Birth of a Star
  - nebulae
    - emission
    - absorption
  - coalescence
  - fusion
- The Sun
  - core
  - photosphere
  - chromosphere
    - spicules
  - corona
  - prominences
- Main Sequence Stars
  - magnitude
    - apparent
    - absolute
  - Hertzsprung-Russell diagram
- Life and Death of a Star
  - sizes
    - Sun-sized

larger sizes
solar generations
Neutron Stars, Pulsars, and Black Holes
neutron star
composition
density
pulsar
black hole
density
Galaxies
composition
shapes
elliptical
spiral
barred
Milky Way
size
composition
motion
Quasars
energy emission
distances
Evolution of Galaxies
hypotheses

Checking Vocabulary

Write a definition for each of the Key Terms at the end of this chapter. Check your knowledge of these terms by answering the following questions.

1. An extremely dense star in which protons and electrons have united is called a
   a. red giant
   b. planetary nebula
   c. protostar
   d. neutron star

2. A light year is the
   a. apparent shift in position of a star in a six month interval
   b. distance to a star with a parallax angle of 1 arc second
   c. amount of time it takes for the sun's radiation to reach the earth
   d. the distance light travels in 365 1/4 days

3. Concentrations of dim and old stars within the galactic halo are called
   a. superclusters

b. dark matter
c. quasars
d. globular clusters

4. A type of galaxy in which the stars are arranged in a thin disk with arms radiating outward from a spherical center is
a. spiral
b. barred spiral
c. elliptical
d. irregular

5. A supernova
a. is produced from an explosion which occurs when ion fusion begins in a star
b. emits radio signals at irregular intervals
c. is produced immediately after hydrogen fusion ceases in a star's core
d. appears as a giant smoke ring in space

6. Dark mater is not
a. an important factor in theories about the future of the universe
b. detectable by modern technologies
c. possibly responsible for the heterogenous nature of the universe
d. a factor which effects the gravity of the universe

7. The idea that most distant galaxies have the greatest red shifts because they are moving outward at the greatest speeds is
a. the Lighthouse theory
b. the Doppler effect
c. Hubble's Law
d. Einstein's theory of relativity

8. The electron force which determines the ultimate diameter of a white dwarf is the
a. degeneracy pressure
b. gravitational force
c. Doppler effect
d. confining pressure

9. The layer of the sun that is visible from the earth under ordinary circumstances is the
a. core
b. photosphere
c. corona
d. chromosphere

10. The spherical arrangement of dust and gas around the Milky Way is called the

a. globular cluster
b. supercluster
c. galactic halo
d. corona

11. Very distant objects which are about four light years across and emit enormous amounts of energy received as radio waves on earth are
a. neutron stars
b. pulsars
c. black holes
d. quasars

12. The stage of star development before nuclear reactions begin is called
a. protostar
b. white dwarf
c. pulsar
d. planetary nebula

13. The Milky Way is the type of galaxy classified as
a. spiral
b. barred spiral
c. elliptical
d. irregular

14. The lighthouse analogy explains regularly spaced radio signals emissions from
a. black holes
b. white dwarfs
c. neutron stars
d. population I stars

15. Galaxies are arranged in groups called
a. globular clusters
b. superclusters
c. galactic haloes
d. main sequence galaxies

Reviewing the Facts

Select the letter of the choice which most correctly completes each of the following statements:

1. The sun's magnetic fields are not associated with
a. sunspots
b. prominences

c. solar flares
d. nuclear reactions

2. Black holes are formed from stars that are
a. less than one solar mass
b. 1.44 to 4 solar masses
c. 4 to 5 solar masses
d. more than 5 solar masses

3. A cool star is
a. white
b. blue
c. yellow
d. red

4. Quasars emit
a. radiation of many wavelengths
b. long wavelength radiation only
c. short wavelength radiation only
d. no radiation

5. The sun is composed mostly of
a. helium
b. hydrogen
c. heavy elements
d. carbon

6. The shape of the galactic halo suggests that the protogalaxy of the Milky Way was
a. elliptical
b. irregular
c. spiral
d. spherical

7. When a star's nuclear reactions stops, its
a. core contracts
b. core cools
c. surface no longer emits light
d. surface contracts

8. The Big Bang theory fails to account for the
a. motion of celestial objects
b. structure of the universe
c. development of energy and matter in the early universe

d. formation of population I stars

9. Main sequence stars located in the upper left of the H-R diagram are
   a. more massive than main sequence stars in the lower right
   b. less bright than main sequence stars in the lower right
   c. cooler than main sequence stars in the lower right
   d. closer to earth than main sequence stars in the lower right

10. Black holes are detected because they
   a. are large
   b. emit radiation
   c. interact with orbiting stars
   d. have infinite density

11. The sun is not expected to become a
   a. neutron star
   b. white dwarf
   c. planetary nebula
   d. red giant

12. Most of the galaxies in the universe are
   a. spiral
   b. barred spiral
   c. elliptical
   d. irregular

13. The sun's total life expectancy is
   a. 1 billion years
   b. 5 billion years
   c. 10 billion years
   d. 15 billion years

14. The main sequence of the H-R diagram does not include
   a. 90% of all known stars
   b. the sun
   c. population I stars
   d. white dwarfs

15. According to the lighthouse analogy, pulsars must
   a. be small
   b. rotate slowly
   c. be close to earth
   d. have a low density

Developing and Applying Concepts

1. Contrast a light year and a parsec.

2. Explain why parallax is not useful in determining the distances to very distant stars.

3. The Greeks rejected a sun-centered model of the solar system because they predicted that parallax should be observed. Why do you think they were unable to verify the existence of parallax in star observations?

4. Match the sun's layer in column A with its corresponding feature in column B:

Column A	Column B
1. core	a. sunspots
2. photosphere	b. spicules
3. outer opaque region	c. prominences
4. chromosphere	d. nuclear reactions
5. corona	e. convection

5. Arrange the following in order of increasing temperature
    1) sun's core
    2) photosphere
    3) sunspot
    4) corona

6. Why isn't the brightness of a star as it appears from earth a true indication of the amount of energy it emits?

7. Since the nuclear reaction of stars in the upper left of the main sequence proceeds at a faster rate, they are only on the main sequence for a short time. Where on the man sequence do you think population I stars are located? Why?

8. Arrange the following in a sequence representing the life history of a star:
    1) supernova
    2) hydrogen fusion in core
    3) contracting nebula
    4) formation of a red giant
    5) protostar
    6) helium fusion in core
    7) neutron star
    8) collapse of core

9. Compare and contrast a globular cluster and a galaxy.

10. Contrast the motion within a nucleus of a spiral galaxy with that of its arms.

Thinking Critically about the Environment

It has been suggested that one way to rid the earth of radioactive wastes would be to put it in a rocket and shoot it into the sun. Find out how much radioactive waste the United States has accumulated. Using modern technologies, determine how many rocket loads would be necessary to rid this country of all its radioactive wastes. Using existing launch costs, attempt to estimate the total cost.

Answers to Questions

Checking Vocabulary

1. d, 2. d, 3. d, 4. a, 5. a, 6. b, 7. c, 8. a, 9. b, 10. c, 11. d, 12. a, 13. a, 14. c, 15. b

Reviewing the Facts

1. d, 2. d, 3. d, 4. a, 5. b, 6. d, 7. a, 8. b, 9. a, 10. c, 11. a, 12. c, 13. c, 14. d, 15. a

Developing and Applying Concepts

1. A light year is the distance that light can travel during one year. One light year equals 9.5×10^{12} kilometers. A parsec is the distance to a star that has a parallax of one arc second. One parsec is equal to 3.2 light years or 3.1×10^{13} kilometers.

2. The farther away a star is, the smaller its angle of parallax. Stars which are very distant have parallax angles that are so small they can't be detected with current technologies.

3. The Greeks did not have telescopes making it impossible for them to detect the parallax angles of stars. The star closest to earth has a parallax angle of 1/4600 of a degree.

4. 1. d, 2. a, 3. e, 4. b, 5. c

5. 1 (sun's core), 4 (corona), 2 (photosphere), 3 (sunspot)

6. Stars are different distances from the earth. Their brightness in the sky is a function of both their distance and the amount of energy they give off. A star that is close to earth will appear brighter than one that is farther away if they both give off the same amount of energy.

7. Population I stars are in the lower right of the main sequence. They have so little mass, they have used hydrogen at very slow rates and they have, as a result, existed a long time. Their brightnesses are low because they are very small and have low temperatures.

8. 3, 5, 2, 8, 4, 6, 1, 7

9. Both globular clusters and galaxies are clusters of stars. Globular clusters tend to be small (10,000 to 1,000,000 stars), spherical in shape, and contain old stars. Globular clusters form a halo around a galaxy. Galaxies are large (billions of stars), contain a higher percentage of young stars, and can be spiral, elliptical, irregular, or barred spiral in shape.

10. X-ray and gamma ray emissions indicate that the matter of the Milky Way's nucleus is moving toward the center of the nucleus. The stars of the arms rotate around the nucleus.